The NEW GENETICS

LEON JAROFF

THE GRAND ROUNDS PRESS

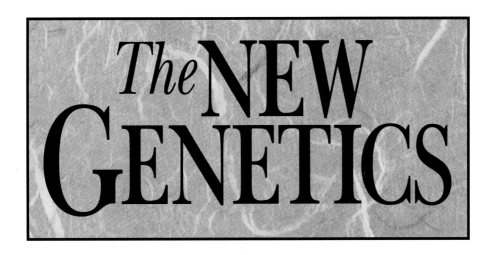

LEON JAROFF

WHITTLE DIRECT BOOKS

Photographs: Gregor Mendel: Brown Brothers, page 10; Thomas Hunt Morgan: the Bettmann Archive, page 11; James Watson: Peter Menzel ©1989, page 15; Ray White: Jeffery Newbury, page 22; Nancy Wexler: Chriss Wade, page 26; Dr. Francis Collins: Pauline Lubens/*People Weekly*, page 34; Dr. Victor McKusick: Manuello Paganelli, page 43; Human karyotype: CNRI/Science Photo Library, page 50; DNA fingerprint: Rainbow/Tom Broker ©, page 62; Dr. W. French Anderson: Manuello Paganelli, page 68; Dr. Steven Rosenberg: Walter P. Calahan ©1989, page 70.

Illustrations by Leonard D. Dank ©1991. Human genome illustration, page 50, adapted from *Mendelian Inheritance in Man*, Dr. Victor McKusick, editor.

Library of Congress Catalog Card Number: 90-071704
Jaroff, Leon
The New Genetics
ISBN 0-9624745-7-6
ISSN 1053-6620

The Grand Rounds Press

The Grand Rounds Press presents original short books by distinguished authors on subjects of importance to the medical profession.

The series is edited and published by Whittle Direct Books, a business unit of Whittle Communications L.P. A new book will be published approximately every three months. The series will reflect a broad spectrum of responsible opinions. In each book the opinions expressed are those of the author, not the publisher or advertiser.

We welcome your comments on this unique endeavor.

PROLOGUE

It was a mild March evening in 1953, and the Eagle, a pub near Cambridge University in England, was crowded with students and teachers unwinding after hours. Oblivious to the din, Francis Crick, 35, and James Watson, 24, raised their glasses in celebration, excitedly recounting what they had just achieved at the university's Cavendish Laboratory. Suddenly, by chance, the hubbub around them waned, and Crick's booming voice could be heard throughout the room. "We have discovered the secret of life," he declared.

Indeed they had. By finally discerning the complex structure of deoxyribonucleic acid (DNA), the giant molecule of heredity, Crick and Watson became the founding fathers of the new molecular genetics. Their discovery resulted in the greatest leap in human understanding of the workings of heredity since Gregor Mendel formulated his laws in 1865. Following that triumph by Watson and Crick, other scientists, building on the knowledge of DNA's structure, have cracked the genetic code, unraveled the incredible workings of the living cell, begun to identify and locate specific genes, and learned to transfer those genes from one organism to another.

Today, most appropriately under the direction of James Watson, scientists have embarked on a monumental effort that could rival in scope—and perhaps exceed in importance—both the wartime Manhattan Project and the Apollo moon-landing program. Over the next 15 years, at an estimated cost of $3 billion, they hope to map all of the estimated 50,000 to 100,000 human genes and spell out the entire message conveyed by three billion chemical code letters in the human genome. That message, written in the code of DNA, is crammed into the nucleus of each of the human body's 10 trillion cells (except red blood cells, which have no nucleus). Its instructions not only determine the structure, size, coloring, and other physical attributes of each human being but can also affect intelligence, susceptibility to disease, and even behavior.

Discoveries made in the course of this ambitious undertaking will have a profound impact on the practice of medicine. To better serve their patients and their own interests, physicians will have to become increasingly knowledgeable about the mechanics of heredity. That will enable them to take full advantage of new techniques for predicting and detecting early in the fetal stages of life inherited diseases like cystic fibrosis, neurofibromatosis, and familial Alzheimer's. Soon doctors will also have the means to gauge susceptibility to many of the more common illnesses, such as cancer and heart disease, that have large genetic components.

As a result, the emphasis in medicine will shift dramatically from the curative to the preventive, and doctors will increasingly prescribe measures and medicines to help ward off disease. For example, high-fiber, lowfat diets and perhaps even cholesterol-lowering drugs may be recommended early in life for those whose genes make them particularly prone to atherosclerosis. Where prevention fails, new drugs and techniques, designed to compensate for such genetic defects as a lack of vital enzymes, will be available. And as scientists become more adroit at manipulating DNA, gene therapy—already a reality—will assume an ever more significant role in medicine.

For all its potential blessings, the new genetics is already raising a host of ethical and legal issues. Does genetic testing constitute an invasion of privacy, for example, and could it lead to discrimination against those with serious inborn deficiencies and further inflame the abortion debate? Will the proliferation of genetic tests make doctors more vulnerable to malpractice and wrongful-life suits? Would knowledge of the human genome be used by employers and insurance companies to identify people who might be occupational or insurance risks?

As the Human Genome Project gathers momentum, these questions are among the issues raised by critics of the project, who are also apprehensive about its "big science" approach and fear that it will divert scarce funds from small research groups. James Watson concedes that problems and doubts exist, but he is confident that they can be resolved and insists that they pale in comparison with the enormous benefits the project will bring. "How can we *not* do it?" he asks. "We used to think that our fate was in our stars. Now we know that, in large measure, our fate is in our genes."

CHAPTER ONE

DOCTORS' DELIGHT, AND DILEMMA

Right now, we wait for people to get sick so we can treat them with surgery or drugs. It's pretty old stuff. Once you can make a profile of a person's genetic predisposition to disease, medicine will finally become predictive and preventive.

—Population geneticist Mark Skolnick

ark Skolnick is hardly a visionary. But he is deeply involved with the future. Working with molecular biologist Ray White at the University of Utah, he has collaborated in mapping the human genome and discovering genetic markers for several hereditary disorders. Like others associated with the Human Genome Project, he is confident that the mapping and sequencing of the genome will lead to a radical change in the way medicine is practiced.

Skolnick is convinced that virtually all disease, even infectious disease, has a significant genetic component. Infectious disease, too? That hardly seems likely. But history, as well as everyday experience, provides some compelling evidence. Skolnick points to the Black Death, which wiped out a substantial portion of Europe's population in the Middle Ages. Why did some people die of bubonic plague while others were unaffected? The answer, says Skolnick, is that for millions of people, "the genetic response of their immune systems wasn't adequate." He concedes that there may be diseases in which the genetic component is close to zero—cancers that would result, for example, "if you inhaled asbestos constantly or had a ball of uranium under your bed." But for most diseases, Skolnick says, that is not the case.

Identifying, mapping, and sequencing the genes that cause disease are major goals of the genome project, but they are only the first steps in the development of tangible medical benefits. Years of additional research will be needed to translate the new knowledge into therapies that can retard or even cure genetic diseases. Still, the mere isolation of genes, or markers close to them, is already having an impact on medicine in the form of accurate diagnostic tests for disease genes.

Although the gene for Huntington's disease has not yet been isolated and the nature of its defective protein remains unknown, for example, the discovery of markers for that deadly gene has led to the development of a Huntington's test that is 96 percent accurate. And only weeks after researchers announced in 1989 that they had identified the cystic fibrosis gene and discovered the mutation in it that causes the disorder in 75 percent of the victims, a biotechnology company announced that it had developed a test for the gene. For those three-quarters of the victims, the test is nearly 100 percent accurate. In addition, the recent discovery and isolation of the genes responsible for such disorders as Duchenne muscular dystrophy, retinoblastoma, and neurofibromatosis have paved the way for tests for these and many other previously mysterious genetic diseases.

In the coming years, as medical researchers discover the root, molecular causes of various disorders, novel preventive strategies, new drugs, and eventually gene therapy will change the face of medicine. While physicians will continue to use traditional methods of diagnosis and treatment, genetics will play an ever-increasing role in medicine, curing or even preventing previously incurable diseases. The newly emerging discipline of genetic counseling will become commonplace.

Biologist Leroy Hood of the California Institute of Technology envisions vast medical potential in the ability, already possessed by researchers, "to look at DNA and determine whether a gene is normal or abnormal." He sees the time in 15 or 20 years when DNA will be extracted from a blood sample taken from every newborn infant. Perhaps a hundred genes in that DNA will automatically be analyzed for flaws that are prediagnostic of disease. "You'll put that information in a computer," Dr. Hood says, "and the computer will give you a printout of the potential life history of that individual with respect to those hundred or so diseases."

The computer will not stop there, says Dr. Hood. "You'll also get a printout that says, 'You should avoid overexposure to the sun. This is the kind of diet you should begin at age 17. These are the kinds of immune-system cells that you should delete at 35. At 50, this is something else you should worry about.' The focus of medicine, basically, will be on keeping people well, on making predictions about the potential difficulties people will run into and then avoiding those problems by manipulating their diet, their environment, or their immune systems, or applying molecular pharmacology."

Long before such automated multidisease screening is achieved, says geneticist Aubrey Milunsky, director of Boston University's Center for Human Genetics, simple, specific diagnostic tests should be available for a wide array of genetic diseases. Once a disease gene has been identified and sequenced, its faulty protein product can be deduced and synthesized. At that point, says Dr. Milunsky, "you don't need any further DNA analysis. You can do a simple immunoassay, just as you do a pregnancy test now."

He adds, "That synthesized protein can be injected into a guinea pig. The guinea pig's immune system recognizes it as a foreign protein and produces antibodies against it." Those antibodies, tagged radioactively, will seek out, cling to, and thereby identify any of that faulty protein in a human blood sample. If the faulty protein is present, the doctor can confirm that the patient carries the disease gene—even if no symptoms are evident—and risks passing it on to any offspring. "Eventually," says Dr. Milunsky, "it will be a spot test in the doctor's office."

That day is drawing closer. Over the past few years, hardly a week has passed without reports about the discovery of a new disease marker, the isolation of an important gene, or the marketing of a newly developed test for a disease gene. And, as the Human Genome Project swings into high gear, it promises to generate new developments in genetics at an ever more dizzying pace.

"We're going to have a hundredfold growth in knowledge of the human genome in the next 15 years," says human geneticist Thomas Caskey of the Baylor College of Medicine. "We in medicine have never had that kind of surge of information." The number of genes and gene markers being identified and located, he says, "is climbing incredibly fast: recessive oncogenes, genes for retinoblastoma and colon cancer. We don't have our hands on all this yet, but we can see it coming."

Dr. Caskey predicts that as newly discovered disease genes are analyzed and sequenced, a host of tests for genetic disorders will quickly follow. Physicians are already feeling the impact, he says, as genes for the common hereditary diseases are identified and more and more patients ask to be tested. Dr. Caskey points to the sudden demand spawned by widespread press reports about the discovery of the cystic fibrosis gene, carried by one in every 25 American whites. His lab, he says, "is being flooded with requests for cystic fibrosis testing."

How will doctors respond to the growing demand? Dr. Caskey is not very sanguine. "The medical community is not well prepared to handle the background information," he says. "The knowledge gap is great." Nancy Wexler, a neuropsychologist at Columbia University who chairs the Human Genome Project's ethics subcommittee, agrees. She tells of one doctor, for instance, who informed a father with a dominant disease that half his children would inherit the gene and would therefore come down with the disease. "He misinterpreted the meaning of risk," she explains. "A 50 percent risk is like flipping a coin. You either have it or you don't. Black or white."

Part of the problem, says Dr. Milunsky, is that "the vast majority of physicians in practice today have never had formal courses or formal training in genetics." As a result, many doctors are uncertain about when to recommend genetic testing or even when to refer patients to medical geneticists. The problem is compounded by a shortage of these specialists. And this, too, says Wexler, "is something that medical schools are going to have to start considering."

Even if all medical schools were to begin concentrating on genetics tomorrow, the dilemma would remain for many doctors in practice today. Dr. Milunsky cites a whole series of genetics issues over which doctors have been sued: failure to do a test, failure to offer a test, failure to inform, providing wrong risk information, saying a disease is not genetic when it is, not knowing that you can detect a disorder prenatally.

In general, such lawsuits charge "wrongful birth," if not malpractice, and are brought by parents of infants born with serious defects. And recently, people with debilitating genetic defects have themselves been suing doctors for malpractice in "wrongful life" actions. They argue, in effect, that life with their defects is worse than no life at all, and that their lives, as well as their suffering, would have been avoided if their parents had received proper genetic advice. Courts in California, Washington, and New Jersey have ruled that such claims are valid.

In his book *Choices, Not Chances* (Little, Brown & Company, 1989), Dr. Milunsky describes a dozen cases in which physicians who gave faulty genetic advice or failed to provide proper advice have been sued, usually successfully, for wrongful birth or malpractice. He cites a typical case:

A New York couple had a child with polycystic kidney disease. The child died five hours after birth. Following autopsy, the parents were reassured that their risks were not increased for having a similarly affected child in a future pregnancy. Litigation began after their second child was born with the same disorder and later succumbed at 2 years of age. Their physician had failed to recognize that this particular kidney disorder was inherited as an autosomal recessive condition and that their risk in each subsequent pregnancy was 25 percent.

Although this doctor was acquitted on a technicality, others have not been so fortunate. Their experiences make it clear that, simply to protect themselves, physicians will have to become knowledgeable enough about genetics at least to decide when to refer a patient to a medical geneticist.

On the other hand, many malpractice suits have been wrongfully brought against doctors who provided proper genetic advice and were ignored or misunderstood. Some states have already acted to protect physicians from this kind of harassment. In 1986, California became the first state to require doctors to offer all pregnant women the fetal test for neural tube defects (including spina bifida and anencephaly). A woman has the right to refuse the test, but if she does so, she is required to sign a waiver of liability that protects the attending physician from legal action resulting from the birth of an infant with any of those defects. In the majority of cases involving genetic advice, however, the legal burden remains on physicians.

Nancy Wexler suggests that one way doctors can meet the new genetic challenges is to concentrate more on their patients' family pedigrees. She quotes a physician friend who says, "If you want to make a diagnosis and you have 10 minutes with a patient, spend seven minutes on the family history and three minutes examining the patient. The family history guides the exam."

When inquiring about familial diseases, says Wexler, it is important for doctors to use terminology that patients understand: "They should try to describe the symptoms of the disease, rather than using labels that patients may not be familiar with." She suggests, for instance, that instead of merely asking, "Did anyone in the family have Alzheimer's?" they should describe specific symptoms: "Was there anybody who was kind of forgetful, who couldn't figure out how to cook dinner, who was always losing her pocketbook?" In Wexler's experience, questions like those often evoke a sudden glint of recognition and a response such as, "Oh, yeah, Aunt Millie!"

Considering a family's risk factors will also change the way doctors practice. "I think there's much more that they can get out of family histories," says Neil Holtzman, a professor of pediatrics at Johns Hopkins University School of Medicine. "In dealing with a relatively young patient who's had a heart attack, the physician will obtain a serum cholesterol level. But even if the level is elevated, I doubt that the internist would recommend the test for the patient's brother, sister, or grown children, though their heart-attack risk is also increased."

In fact, Dr. Holtzman says, "genetic tests may be more effective than

7

lipid-level tests for predicting whether relatives are at risk or not, though few such tests are now available." But physicians in general need to be aware of genetic testing, he adds, and especially in cases involving reproduction, should know when to suggest them.

When does it become the responsibility of a doctor to recommend a specific test? "If a test becomes the standard of care," says Dr. Holtzman, "failure to offer it to a patient at high risk for that genetic disease makes a doctor vulnerable to a malpractice or wrongful-birth suit." And when does a particular genetic test become the standard of care? "That is not clearly defined," says Dr. Holtzman, "but generally it happens when professional societies or consensus conferences deem it to be the standard, when a state makes the test mandatory, or simply when most doctors accept it and begin to use it."

The newly developed test for cystic fibrosis, for example, is accurate only for the 75 percent who carry the CF gene with the most common mutation. "There are still too many unknowns and uncertainties in the test," says Dr. Holtzman. "Among couples who are tested for the CF mutation, there will be many in which one partner is found to be a carrier but the other partner's carrier status cannot be determined." Consequently, at its December 1989 meeting in Baltimore, the American Society of Human Genetics declared that the CF test was not yet the standard of care. Subsequent work has shown that there are at least 60 different mutations in the CF gene, dimming the prospects for the speedy development of a single, definitive cystic fibrosis test. Even the Huntington's test has not yet become the standard of care.

Unfortunately, the genetic-testing area can still be a minefield for doctors because the judgment of a court, and not necessarily the considered opinion of an association, may prevail. Dr. Holtzman gives an example: "If the parent of a kid with CF goes to court and says, 'This test was available and the doctor didn't provide us this information, even though it wasn't the standard of care,' the judge may say, 'Well, if this test was available, it was prudent for the doctor to order it.'" Aware of that possibility, many physicians have begun practicing defensive medicine, ordering a test even when, in their judgment, it is uncalled for—simply to ensure that they have covered all legal bases.

Despite the additional burdens and potential pitfalls the new genetic discoveries bring, their promise far outweighs any temporary concern they may arouse in doctors, their patients, and society. The fact remains that help is on the way to people with serious genetic disorders, and the Human Genome Project is speeding its arrival.

THEN AND NOW

▼

In 1956, Squibb Diagnostics introduced the first iodinated contrast
medium: Renografin® (diatrizoate
acid meglumine/sodium). In 1986,
Squibb launched Isovue® (iopamidol injection), the nonionic with the
best iodine-to-carrier ratio. In 1991,
Squibb Diagnostics' tradition of research continues with new agents
being developed for CT, MRI, and
nuclear medicine.

ISOVUE® ...*enhances your image*
(iopamidol injection)

IN LINE

Right now, new nonionic contrast media and new imaging agents for MRI and nuclear medicine are moving through the R&D pipeline at Squibb Diagnostics.

SQUIBB™
Diagnostics

A Bristol-Myers Squibb Company

3/91

CHAPTER TWO

THE TALMUD, FRUIT FLIES, AND GENES

From the very beginnings of history, human beings have been fascinated by the distinctive traits that run in families generation after generation—particular facial features, hair coloring, physical deformities, susceptibilities to disease, and even mental disorders. Ancient civilizations, recognizing this basic principle of heredity, used it in breeding animals and plants to improve their characteristics.

Other early genetic insights were written into the Talmud, the ancient Hebrew compendium of civil and criminal law. Biologist Eric Lander of the Whitehead Institute for Biomedical Research in Cambridge, Massachusetts, points out a Talmudic provision that exempts a Jewish boy from circumcision if a maternal uncle (but not a paternal one) is a hemophiliac. That exception, says Lander, "reveals a sophisticated understanding that this sex-linked trait is inherited only through the mother."

But how are these characteristics passed from generation to generation? And why do children inherit some ancestral features but not others? Addressing themselves to the phenomenon of heredity in the third century B.C., the philosopher Aristotle and other ancient Greeks proposed the theory of pangenesis—widely accepted for nearly two thousand years—that traits are inherited through the blood. The residue of Aristotelian theory persists today in such words as "blood relatives" and "bloodline."

That theory was still largely unchallenged in the late 1600s when, after the invention of the microscope, scientists first observed the human sperm cell. This discovery prompted some 17th-century biologists to propose that sperm cells contained a tiny, fully formed person, or homunculus, that grew to full size in the mother's womb. This "preformation" concept remained conventional wisdom until the late 18th century, when German scientist Kaspar Wolff proposed the theory of epigenesis, which held that the fetus develops through a process of differentiation as well as growth.

But it was not until 1865 that the Austrian monk Gregor Mendel applied scientific methodology to the enigma of heredity and proposed that the mysterious process was governed by a set of rather simple laws. Working with pea plants in the garden of his monastery, he demonstrated that specific traits—flower color, pod shape, seed color, plant height—are passed down from one generation to another in a mathematically predictable manner.

Gregor Mendel, the Austrian monk whose observations of inherited traits in peas in 1865 led to the first scientific laws of genetics.

From the results of these experiments, he proposed the three-part law of segregation:

- The traits of heredity are carried in discrete "factors" (subsequently named genes), which occur in pairs.
- Each pair of factors separates when the sex cells are formed, and each sperm or egg receives only one member of a factor pair.
- Each factor in a pair will be found in half the sperm and egg cells.

Mendel further concluded that if one factor in a pair is dominant and the other recessive, the dominant trait is always expressed. If both factors in a pair are recessive and identical, the recessive trait will be expressed.

Although Mendel published the details of his work and his laws in 1866, his report was largely ignored for decades. But other scientists continued to probe the mechanisms of the living cell. In 1877, German scientist Walther Flemming detected chromosomes and other scientists distinguished a chromosomal difference between the sexes: males had one X and one Y chromosome, while females had two X chromosomes. All the remaining chromosomes were paired and seemed identical in males and females.

By the turn of the century, biologists had discovered that the germ cells (the sperm and egg cells) are produced in a process now called meiosis, from a special class of progenitor cells. During meiosis the chromosomes in these cells fragment, each pair exchanging segments, and reconstitute themselves in a process called crossing over, or recombination. The pairs then separate, leaving each of the resulting egg or sperm cells with just half of the normal cell's chromosomes. When fertilization occurs, the chromosomes in the sperm and those in the egg pair up to restore the full complement of chromosomes in the fertilized egg.

These observations more or less coincided with the rediscovery of Mendel's work by European biologists. A Columbia University graduate student, Walter Sutton, promptly made the connection. In 1902, he called attention to the parallelism between Mendel's factors and the behavior of chromosomes during meiosis. The conclusion was obvious: Mendel's factors, or genes, as they were soon afterward named, must reside in the chromosomes.

This confluence of Mendelian theory and laboratory evidence marked the beginning of modern genetics. At Columbia University, Thomas Hunt Morgan and his colleagues, working with fruit flies, discovered several important hereditary mechanisms. Among them were sex-linked genes and traits that appeared to be linked to each other and therefore co-inherited.

Thomas Hunt Morgan, with colleagues from Columbia University, produced a crude genetic map during the first decade of the 20th century.

This linkage of traits, Morgan assumed, indicated that genes for the two traits might lie close together on a chromosome. That meant that during meiosis, when segments of the homologous chromosomes paired, they would cross over and recombine. Closely linked genes would probably not be separated and would be inherited together. Conversely, he reasoned, if two traits were seldom co-inherited, they must lie far apart on a chromosome or on different chromosomes.

"That was a very clever insight," says Norton Zinder, the Rockefeller University biologist who heads the Genome Advisory Committee. "The distance between two genes was reflected in the percentage of recombination that you got. When you got more recombination between two genes—in other words, when they were seldom co-inherited—they were further apart. When there were fewer recombinants between two genes and they were frequently co-inherited, they were closer together."

Using selected matings and linked traits, and deducing the approximate location on the X chromosome of sex-linked genes for eye color, wing shape, and other obvious traits, Morgan and his co-workers produced the first crude genetic map. Their so-called linkage map placed genes for many fruit fly traits in linear order along the X chromosome.

In 1911 the first human gene was correctly assigned to a particular chromosome. After studying the pedigrees of several large families with many colorblind members, geneticist E. B. Wilson, also at Columbia University, recognized that although the disorder was primarily a male affliction, it was passed from generation to generation by women who had normal color vision. Applying Mendelian logic and building on Morgan's work with fruit flies, Wilson assigned the gene to the

X chromosome. He reasoned that women, who have two copies of the X chromosome, are protected against color blindness by a gene in one X chromosome even when its counterpart gene in the other X chromosome is defective. When men inherit the defective gene on their sole X chromosome, however, there is no corresponding normal gene to protect them.

In the same manner over the next few decades, several genes responsible for such gender-linked diseases as hemophilia were assigned to the X chromosome (providing confirmation, of sorts, for the ancient Talmudic exemption).

Scientists remained uncertain about the exact number of human chromosomes until 1956, when photomicrographs of dividing cells established that there were 46 chromosomes within the human cell nucleus—one pair of sex-determining chromosomes (the two X's in females, the X and Y in males) and 22 sets of non-sex chromosomes (called autosomal chromosomes) that are identical in both sexes. This revelation led directly to identifying the cause of Down syndrome— three rather than the normal two copies of chromosome 21.

Scientists were now faced with a greater challenge. How could they assign a particular gene to any of the autosomal chromosomes? Part of the answer had been provided in 1941 by biologists George Beadle and Edward Tatum. Working with a type of mold called *Neurospora*, the American scientists discovered that mutations in a gene always caused abnormalities in only one protein and that for a given mutated gene, the defect always occurred in the same protein. Apparently, the scientists proposed, each gene governs the formation of a single, specific protein.

That conclusion had been confirmed and the cell's protein-building process elucidated by the early 1960s, when researchers learned to fuse human skin cells with mouse cells. The technique (which the British biologist J. B. S. Haldane, tongue in cheek, called "an alternative to sex") produced hybrid cells that contained both mouse and human chromosomes. But when these cells were placed in a nutrient-filled laboratory dish to divide and multiply, they gradually shed their human chromosomes.

The discriminatory behavior of the hybrid cells gave the researchers an idea. Since each gene orders the production of a single, distinctive protein, the scientists decided to analyze the proteins produced by hybrid cells that had shed all but one, or even a fragment of one, human chromosome. Then, by identifying the human protein in the cellular products, they could deduce that the gene responsible for that protein resided in the surviving human chromosome or fragment. Using this ingenious technique, known as forward genetics, researchers have now assigned hundreds of genes to specific chromosomes, and in many cases to a particular segment of a chromosome.

They were aided in their gene-mapping efforts by the discovery, in 1969, that chemically stained chromosomes revealed alternately dark and light banding patterns and that each set of chromosomes had different and unique band characteristics. The patterns enabled researchers not only to distinguish one human chromosome from another (several of the chromosomes have nearly identical shapes) but to identify the human chromosomes in a hybrid mouse-human cell. The bands also provided a kind of grid on which mappers could more accurately place specific genes, and made it easier to distinguish which segments of the chromosomes were exchanged during meiosis.

By the mid-1970s, scores of genes had been assigned to the X chromosome, at least one gene had been tracked down to the Y chromosome, and dozens had been traced to the autosomal chromosomes. In practically every case, researchers were able to locate the autosomal gene by working backward from the protein associated with it. But there were tens of thousands of other genes—including those that caused disorders such as Huntington's disease and cystic fibrosis—that could not be located because the protein they "coded" for had not been identified. And it was becoming increasingly apparent that little more progress could be made without some radically new ways of probing the secrets of chromosomes at the molecular level.

It was at this point that a powerful, new technique—recombinant DNA technology—was coming to the fore. It was to open vast new areas in genetics research and eventually make possible the conception and launch of the Human Genome Project.

CHAPTER THREE

DOUBLE HELIXES, SCISSORS, AND RFLPS

For all the progress scientists had made by 1940 in probing the mysteries of heredity, they had few clues about the nature of Mendel's "factors," the genes, or how they performed their prodigious feats. How, for example, did they store what must certainly be a huge volume of information about a living organism? And how did they use that information not only to sustain the life of the organism but to faithfully reproduce its traits in succeeding generations?

The 1941 discovery by Tatum and Beadle that each gene controls the production of a specific protein shed some light on the mystery, but left scientists in the dark about what a gene actually was. The answer obviously lay in the chromosomes, which were known to consist of both protein and a chemical identified as deoxyribonucleic acid, or DNA. Somehow, it was generally agreed, the message of the genes must be written in the complex molecular structure of one or both of these substances.

In 1944, geneticist Oswald Avery and his co-workers at the Rockefeller Institute identified the message carrier. They extracted DNA from one type of bacteria, purified it, and introduced it into the cells of a different variety of bacteria. When the bacteria in the second group divided and reproduced, their progeny carried new traits that were characteristic of the cells in the first group. It was obvious to Avery that new genes had been incorporated into the second group of bacteria. And since only purified DNA had been transferred between the groups, he reasoned, genes must consist of DNA.

James Watson, co-discoverer with Francis Crick of the structure of the DNA molecule, now directs the Human Genome Project.

Word of Avery's experiment was greeted with some skepticism, and his conclusion was not generally accepted until 1952, when Alfred D. Hershey and Martha Chase, at the Cold Spring Harbor Laboratory in New York, used radioactive labeling in an elaborate experiment to prove that DNA was indeed the stuff of heredity. Their achievement, which eventually won Hershey a share of the 1969 Nobel Prize in Physiology or Medicine, sparked frenzied research in laboratories around the world as scientists competed to discover the structure of DNA and learn how the giant molecule stored and conveyed the voluminous message of heredity. Among those who joined the fray was James Watson, who in his remarkably candid book, *The Double*

Helix (Atheneum Publishers, 1968), described how he and Francis Crick combined inspiration, conspiracy, perseverance, and cunning to unlock the secret of DNA.

The historic report documenting their discovery was printed in the April 25, 1953, issue of the British journal *Nature*. It occupied little more than a page and was written with understatement and with a clarity rare in scientific literature. "We wish to suggest a structure for the salt of deoxyribose nucleic acid (D.N.A.)," it began. "This structure has novel features which are of considerable biological interest." Novel, indeed.

As Watson and Crick discovered, DNA is wondrously complex. It consists of a double helix that resembles a twisted ladder, with side pieces made of sugar and phosphates and closely spaced connecting rungs. Each rung is called a base pair because it consists of a pair of nitrogenous bases (nucleotides) attached end to end. Only four of those paired nitrogenous bases are found in DNA, either adenine (A) joined to thymine (T) or cytosine (C) attached to guanine (G).

The significance of DNA's molecular structure was immediately obvious to its discoverers. "It has not escaped our notice," they stated in their 1953 report, "that the specific pairing we have postulated immediately suggests a possible copying mechanism for the genetic material."

Fundamental to the genius of DNA is the fact that A and T are mutually attractive, as are C and G. As Watson and Crick suggested, when the DNA "ladder" comes apart during cell division, it separates at the middle of each rung, somewhat like a zipper opening, leaving two separate strands that each carry half-rungs of A, T, C, or G.

These half-rungs are joined by their complementary numbers from the countless nucleotides that float freely in the nucleus of the cell. A floating A will attach to a T half-rung, for example, a G to a C, and so on, until two double helixes are completed. Each consists of a strand from the parent molecule and a newly assembled strand, and both—barring errors in transcription—are identical to the original DNA molecule.

Furthermore, each of the four bases represents a letter in the genetic code. And most of the three-letter words, or codons, that they spell, reading in sequence along either side of the ladder, code for one of the 20 different amino acids that are the building blocks of all proteins. The sequence CAC, for example, specifies the amino acid histidine, and GCA codes for alanine. A few codons represent punctuation, specifying either the beginning or the end of a genetic "sentence." Each of these sentences is a gene, a discrete segment of the DNA string between 10,000 and more than 200,000 code letters long.

Learning how genetic information is stored by DNA was triumph

EDUCATIONAL TELEVISION

▼

Squibb Diagnostics produces in-service films for radiologists, radiology departments and administrators, including *The History and Development of Contrast Media* and *Conversion to Nonionic Contrast Media.*

...*enhances your image*
(iopamidol injection)

PATIENT-WISE

▼

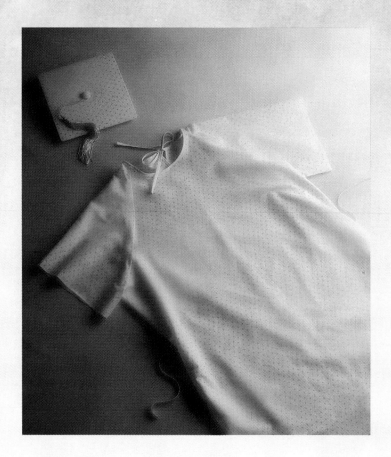

Through easily understood educational videos and brochures produced by Squibb Diagnostics, many patients' fears about upcoming imaging procedures are allayed.

SQUIBB™
Diagnostics

A Bristol-Myers Squibb Company

3/91

enough. But that achievement left scientists with an even greater challenge: to learn how that information was translated into the production and sustenance of living organisms. By the late 1950s, molecular biologists were well on their way toward understanding the astonishingly complicated process of protein-building, helped along by the discovery of the key role played by another large molecule found in living cells: ribonucleic acid, or RNA.

When the time comes for a gene to turn on, or express itself, only the segment of DNA containing that gene unwinds and splits down the middle. (The ensuing steps, resulting in the production of a protein molecule, are shown in the diagram on page 18). If the process goes wrong at any stage, a faulty protein results. In the case of a genetic disorder such as sickle cell anemia, the deletion of just the single nucleotide A—a point mutation—in the gene that controls hemoglobin production results ultimately in the deformed red blood cells characteristic of this disease.

By the late 1960s, scientists had completely deciphered the genetic code and largely elucidated the function of the gene and the cellular processes. As scientists translated more and more of the DNA message, however, they discovered between the genes large sequences of the genome that apparently made no sense.

"My feeling is that there's a lot of very useful information buried in the sequence," says Nobel laureate Paul Berg of the Stanford University Medical Center. "Some of it we will know how to interpret; some we know is going to be gibberish."

In fact, some of the non-gene regions on the genome have been identified as instructions necessary for DNA replication during cell division. Those instructions are obviously detailed and complex. As George Bell, former head of genome studies at Los Alamos National Laboratory, explains, "it's as if you had a rope that was maybe two inches in diameter and 32,000 miles long, all neatly arranged inside a structure the size of a superdome. When the appropriate signal comes, you have to unwind the rope, which consists of two strands, and copy each strand so you end up with two new ropes that again have to fold up. The machinery to do that cannot be trivial." Learning the nature of that machinery and other genetic instructions buried in the long noncoding DNA sequences is one of the long-range goals of the Human Genome Project.

Even more puzzling are the short non-protein-coding segments of DNA, called introns, that are found within genes. As few as two and as many as 50 introns are interspersed between groups of amino-acid-coding words (called exons) in a gene sentence. Scientists can only speculate about the purpose of introns. Caltech's Leroy Hood, for one,

Protein Synthesis

The central dogma of molecular biology is DNA ▶ RNA ▶ protein. DNA contains the information for the linear order of amino acids in a particular protein; RNA carries the instructions encoded in DNA from the cell nucleus to the cytoplasm, where it specifies the kind of protein that the cell machinery will produce.

1 RNA is made from DNA in a process called transcription. Here, an enzyme called RNA polymerase finds an initiation site on DNA, separates the strands, and copies a complementary RNA strand from the DNA template.

2 This pre-messenger RNA consists of start and stop fragments, one at each end, and a succession of introns (noncoding sequences) and exons (expressed, or coding, sequences).

3 After the introns are removed, the separated exons are spliced.

4 The result is messenger RNA, which consists of a series of three-letter code words, each specifying an amino acid.

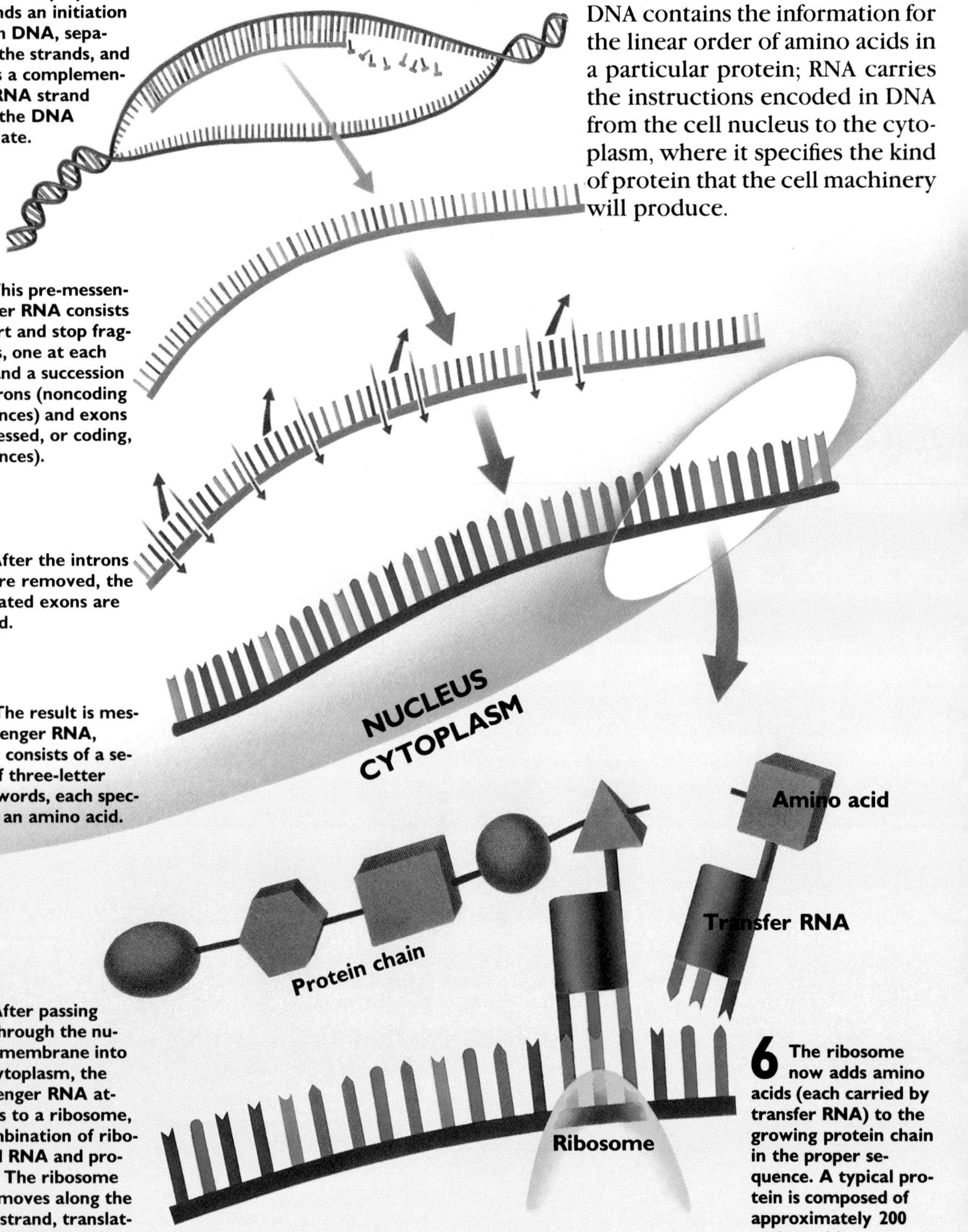

NUCLEUS

CYTOPLASM

Amino acid

Transfer RNA

Protein chain

5 After passing through the nuclear membrane into the cytoplasm, the messenger RNA attaches to a ribosome, a combination of ribosomal RNA and proteins. The ribosome then moves along the RNA strand, translating its coded message.

Ribosome

6 The ribosome now adds amino acids (each carried by transfer RNA) to the growing protein chain in the proper sequence. A typical protein is composed of approximately 200 amino acids.

suggests that primordial cells may have contained chromosomes with both introns and exons. Then, "when you split off to become bacteria," he says, "you were very small and you had to be very efficient, so you clipped out all of your introns so that your genes could all be packed against one another."

For those that became larger, multicellular organisms, however, the introns provided more evolutionary flexibility. "You could shuffle exons around in a way that bacteria couldn't without disrupting or destroying genes," Dr. Hood explains. "So in evolutionary terms, we can put information together in all sorts of different combinations." That capability, he believes, is one of the features that made possible the evolutionary changes that eventually culminated in *Homo sapiens*.

Even as they read and translated the message of DNA, scientists were looking ahead. The next obvious step was to learn how to manipulate the submicroscopic genetic machinery in order to identify and locate genes in the chromosomes, spot any genetic errors, and eventually correct or compensate for them. But how?

The answer was provided in the early 1970s by the discovery of restriction enzymes in bacteria. These enzymes, a form of protein, have the uncanny ability to "recognize" certain sequences of nucleotide code letters and to sever the DNA strand wherever those sequences occur. A restriction enzyme called *Eco* RI, for example, cuts through DNA wherever it encounters the sequence GAATTC, cleaving the strand between the G and the A.

Nearly a hundred of these restriction enzymes have been found in bacteria, each recognizing and cleaving a different DNA sequence. Their purpose, says Dennis Ross, a pathologist at the University of North Carolina at Chapel Hill, is to protect bacteria against viruses. "It is immunology at the lowest level," he says. For example, recognizing the sequence GAATTC in the genome of an invading virus, the bacterial enzyme cleaves the viral DNA, which prevents the virus from replicating and destroying the bacterium. Since the bacterium does not have the GAATTC sequence in its own rather small and simple genome, the enzyme causes it no harm.

Humans do not have restriction enzymes, and for good reason. The human genome is thousands of times more complex than the bacterial genome, and, says Ross, "every restriction enzyme sequence ever discovered occurs in the human genome." Consequently, he explains, "the human genome cannot have restriction enzymes as defense mechanisms because we would be cutting our own DNA in an effort to protect ourselves against viruses."

Discovery of the restriction enzymes gave molecular biologists the tools they had been seeking. The enzymes, says Paul Berg, "provided

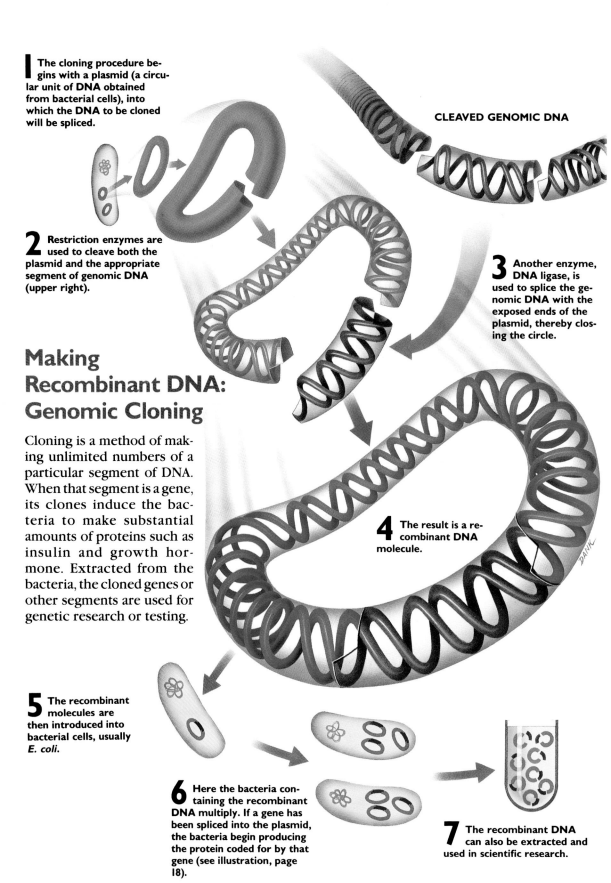

1 The cloning procedure begins with a plasmid (a circular unit of DNA obtained from bacterial cells), into which the DNA to be cloned will be spliced.

CLEAVED GENOMIC DNA

2 Restriction enzymes are used to cleave both the plasmid and the appropriate segment of genomic DNA (upper right).

3 Another enzyme, DNA ligase, is used to splice the genomic DNA with the exposed ends of the plasmid, thereby closing the circle.

Making Recombinant DNA: Genomic Cloning

Cloning is a method of making unlimited numbers of a particular segment of DNA. When that segment is a gene, its clones induce the bacteria to make substantial amounts of proteins such as insulin and growth hormone. Extracted from the bacteria, the cloned genes or other segments are used for genetic research or testing.

4 The result is a recombinant DNA molecule.

5 The recombinant molecules are then introduced into bacterial cells, usually *E. coli.*

6 Here the bacteria containing the recombinant DNA multiply. If a gene has been spliced into the plasmid, the bacteria begin producing the protein coded for by that gene (see illustration, page 18).

7 The recombinant DNA can also be extracted and used in scientific research.

the molecular scissors that enabled us to cut up DNA and to study the arrangement of genes in DNA."

Using those scissors in 1971, Berg and his associates sliced open the rather simple circular genome of simian virus 40, a well-studied monkey virus, and spliced it into the genome of a lambda bacteriophage, a virus that infects bacteria. For this first-ever artificial joining of DNA from different organisms, Berg shared the Nobel Prize in Chemistry in 1980.

The following year two other researchers, Stanley Cohen of Stanford and Herbert Boyer of the University of California at San Francisco, used restriction enzymes to develop a cloning technique (see the diagram on the opposite page) that has been widely used in genetic engineering. For devising this method of making virtually unlimited numbers of a gene or any other segment of DNA, Cohen and Boyer were awarded the first patents in the field of recombinant DNA. Variations of the technique have since been used to mass-produce human insulin, growth hormone, and other pharmaceutical protein products of human genes.

Armed with their recombinant tools, scientists devised new and faster ways to clone segments of DNA, to read the sequence of base pairs in those segments, and to synthesize lengths of DNA from on-the-shelf laboratory chemicals. With these techniques, the gene hunters devised still another way to locate genes. By analyzing the amino-acid sequence in a faulty protein that had been associated with a genetic disease, they could deduce the DNA sequence in the gene responsible for that protein. They then synthesized strands of DNA with that particular sequence, tagged each with a radioactive marker, and used them as probes. Intermingled with "denatured" chromosomes (chemically treated to separate their double helixes into single strands) and propelled by the powerful attraction between DNA nucleotides, the probes sought out and then clung to their complementary sequences on the chromosomal DNA. In a photomicrograph, a probe appeared as a bright spot on one of the dark chromosomes, dramatically marking the location of the gene.

Using this and other methods of forward genetics, by which genes are found through their protein products, researchers were able to locate many human genes. Among them were those responsible for sickle cell anemia, Tay-Sachs disease, hemophilia, thalassemia, and hypercholesterolemia, all of which have been pinned to specific faulty proteins.

Still, for most of the other 3,500 known genetic diseases, the defective protein was unknown. Medical researchers were stumped. How might they even begin to track down a gene responsible for an

Ray White, a molecular biologist at the University of Utah, has created extensive genetic-linkage maps by using the RFLP technique on DNA obtained from large Mormon families who live in the state.

unknown protein, and from that gene discern the nature of the protein defect that causes the disease? The answer lay in "reverse genetics," which University of Michigan geneticist Francis Collins wryly describes as "the business of trying to find a gene when you don't know what it does."

In seeking that answer, scientists recalled the work of Thomas Hunt Morgan in mapping the genes of his fruit flies in the early part of this century (see Chapter Two). After locating single genes that determined eye color, wing shape, and bristle patterns, Morgan had used them as genetic markers to help him locate the position of other genes closely linked to the markers. In the years that followed, other scientists had used similar techniques to find markers and map the genes of yeast, bacteria, flies, and mice.

But constructing such a linkage map in humans presented what seemed to be two insurmountable problems. Eric Lander of the Whitehead Institute explained: "First, human matings could not be arranged to suit the design of the experimenters. Second, almost no 'genetic markers' were known relative to which traits could be mapped." Most human traits like eye color and hair color, Lander said, are not the result of single genes. Consequently, "they cannot serve as markers for particular genetic regions."

As the search for human markers bogged down, however, researchers using restriction enzymes to slice up segments of DNA noticed a phenomenon that was to prove a godsend to gene hunters. When DNA segments from identical places on the chromosomes of different people were sliced by the same restriction enzyme, the lengths of the resulting DNA fragments often differed between individuals. The explanation was rather straightforward: slight differences exist between the nucleotide sequences in human DNA—a missing base pair here, an extra one there. University of Utah molecular biologist Ray White notes that, on average, only one in 500 base pairs will differ from person to person. "You and I," he says, "are 99 percent plus identical at the DNA level."

But when one of those variations occurs at a site that is normally recognized and cut by a restriction enzyme, the enzyme passes it by. For example, the enzyme *Hpa* II always cuts the sequence CCGG. If that sequence occurs in the middle of a DNA segment from one person, the enzyme will cut the segment in two. In the corresponding DNA segment of another individual, however, a C may be missing, and the enzyme will ignore that site, leaving the segment intact. The differences in length of DNA fragments from different people were given the name restriction fragment length polymorphisms (RFLPs), dubbed "riflips" by geneticists. These RFLPs, or variations, had another significant characteristic: they were inherited in families. Although scientists did not immediately recognize it, this characteristic made RFLPs a powerful new tool for hunting elusive genes.

The idea of using RFLPs as signposts, or markers, along human chromosomes was inspired by a lecture at a University of Utah genetics workshop in 1978; it dealt with one of the few easily recognizable markers in the human genome, HLA genes. These genes lie close together in one region of chromosome 6 and are responsible for the production of HLA, an important immune-system protein. The HLA genes are usually co-inherited with a gene that, when defective, causes hemochromatosis. To geneticists, this co-inheritance indicates that the hemochromatosis gene is tightly linked to the HLA genes and therefore lies in the same region of chromosome 6.

At the workshop was David Botstein, then a molecular biologist at MIT, who suddenly had a momentous insight. Could RFLPs in various places along the genome be used as markers for the elusive disease genes whose protein products were unknown? After the workshop session, Botstein engaged in a spirited discussion about the idea with Ron Davis, a molecular biologist at Stanford University. "The more we discussed it," says Davis, "the more it seemed like a right idea."

RFLPs might be ideal as markers, the two biologists decided, because they were often highly polymorphic between individuals, and they were inherited. This meant that they could be tracked from generation to generation in a family. To be valuable in locating genes, moreover, the marker had to sit close enough to the gene of interest that it would not be lost during the recombination that occurs in meiosis. By tagging genes of interest with nearby RFLPs, Botstein and Davis concluded, researchers could finally begin to construct a true genetic-linkage map for humans of the kind already done for the fruit fly and other species.

For example, explains Eric Lander, "suppose that a restriction enzyme cut a grandfather's or grandmother's DNA into slightly different patterns at a particular location. By studying the pattern of fragments in several grandchildren, one could detect which has inherited Grandfather's copy of the region and which has Grandmother's copy. Now, if those grandchildren who inherited a particular genetic disease from Grandfather were always the ones who had inherited a RFLP from Grandfather, then the RFLP must lie very close to the gene for the disease. Eventually, random trial of the RFLPs would uncover one in the right neighborhood."

Botstein's RFLP idea quickly gained support, and in a 1980 paper in the *American Journal of Human Genetics*, Botstein, Davis, Ray White, and Mark Skolnick formally proposed using RFLPs for constructing a genetic-linkage map of the entire human genome. True to their promises, RFLPs were soon playing a key role in one of the more dramatic endeavors in medical history: the successful hunt for the elusive Huntington's gene.

RAISING RADIOLOGY

▼

Squibb Diagnostics is a Vanguard Contributor to the Radiological Society of North America, providing educational support for the next generation of radiologists.

1

ENHANCING PARTNER

Squibb Diagnostics wants to be
your partner in enhancing the
quality of diagnostic medicine.

 **SQUIBB**™
Diagnostics

A Bristol-Myers Squibb Company

Chapter Four

The Great Huntington's Hunt

F or as long as they could remember, the citizens of Maracaibo, Venezuela, had shunned the residents of the villages that line the shore of nearby Lake Maracaibo. They had grown used to seeing the villagers staggering down the street, weaving from side to side in jerky motions, and suspected that they were afflicted with a mysterious disease.

In the early 1950s, when a Venezuelan doctor named Americo Negrette arrived to practice a year of rural service in the area, he at first attributed the bizarre behavior to alcoholism. But it soon dawned on Dr. Negrette that what he was witnessing, on a massive scale, was symptoms of Huntington's disease, a genetic disorder that generally strikes around middle age, causing degeneration of brain cells and eventual loss of control of the voluntary muscles. As the disease progresses, the victims suffer dementia and finally, 10 or 15 years after the first symptoms, death.

Amazed at the remarkably high incidence of what is usually a rare genetic disease, Dr. Negrette conducted a survey of the villagers, sketched out some family trees, and soon had an explanation: virtually all of the villagers belonged to an interrelated set of families with a common heritage. In 1972, at a Columbus, Ohio, meeting of the World Federation of Neurology Research on Huntington's Disease, one of Dr. Negrette's students showed a film of the villagers and presented the physician's report on the case histories and pedigrees of about a hundred of the Huntington's victims.

Among those in the audience were Nancy Wexler and her father,

Nancy Wexler stands beside the pedigree chart of a Venezuelan Huntington's disease family that includes more than 10,000 members. The pedigree has been crucial to the search for the Huntington's gene.

Milton, both clinical psychologists, and Nancy's sister, Alice. All of them had more than an academic interest in the project. Three of Nancy Wexler's uncles on her mother's side had died years earlier of Huntington's disease, which her family mistakenly thought was something that affected only men. But in 1968, her mother began to show the same symptoms, and Wexler learned to her dismay that the disease was caused by a dominant gene.

"That meant that both my older sister and I were at 50 percent risk of developing Huntington's," she says. "We decided to fight." Milton Wexler raised funds and set up the Hereditary Disease Foundation. He served as president for 13 years and was succeeded by his daughter Nancy. The foundation began sponsoring three or four workshops a year, inviting young scientists to brainstorm ideas for detecting and fighting inherited disorders. Though its initial goal was to identify the gene or genes responsible for the disorders, the foundation's ultimate aim was to devise a strategy to circumvent the dire effects of these faulty genes.

One of the workshops was held in conjunction with the 1972 Huntington's meeting in Columbus, where the Wexlers heard Dr. Negrette's report. "We were all very excited," Nancy Wexler recalls. "We thought that this ought to help answer something, but we couldn't figure out the question. This was a potentially fantastic genetic resource—so many patients in the family. But at that point nobody knew how to formulate the right research."

Four years later Michael Brown and Joseph Goldstein published their famous paper on familial hypercholesterolemia, a disorder caused by a dominant gene. In that paper, the two scientists described how they had found that gene, a discovery for which they shared the Nobel Prize in Physiology or Medicine in 1985. As Wexler read their report, she was struck by the fact that they had been helped in their search for the culprit gene by chancing upon a little girl who had inherited a copy of the defective gene from each of her parents and was having heart attacks at the age of 7.

In most cases of a dominant genetic disease, the victim has one normal copy and one abnormal copy of the gene. Because the defective gene is dominant, the disease is present, but its nature and effects may be attenuated by the good gene. In the rare case of a homozygote—someone carrying two copies of a dominant defective gene—the workings of the gene are magnified, and starkly evident. "So if a person doesn't have the normal gene to clutter up the works," Wexler explains, "you might get a clue."

Her thoughts turned immediately to the Venezuelan Huntington's family. Among the hundreds of Huntington's victims around Lake Maracaibo, she figured, there must be a homozygote who might help identify the gene and shed some light on the disease. In 1979, with funds from the National Institutes of Health and the Hereditary Disease Foundation, she led a small team to Lake Maracaibo.

Seeking out families in which both parents had Huntington's, Wexler took blood samples that were returned to the United States for analysis. "In one fantastic family," she says, "we found both parents with Huntington's and 14 children," one of whom was a homozygote. But DNA analysis of his blood failed to turn up the sought-for clue.

Wexler's Hereditary Disease Foundation next enlisted the help of molecular biologist David Housman of MIT, who suggested that RFLPs might expedite the search for the Huntington's gene. "Find a marker that's next to it," he said, "and you've got the gene."

At the foundation's request, Housman organized a Huntington's workshop at the National Institutes of Health in October 1979, inviting Botstein and White, among others. The participants decided that the most direct route to finding the gene was to begin building an extensive pedigree of the interrelated Venezuelan Huntington's families and analyzing the DNA of each person needed for the study. That, at least in theory, would enable scientists to find marker patterns common to all those with the disease, localize the marker to a particular chromosome, and then begin zeroing in on the precise location of the gene.

With that goal in mind, Wexler returned to Lake Maracaibo in March 1981. Her team began a massive effort to interview the villagers, trace family histories, and collect blood samples and skin biopsies.

At first people were embarrassed and reluctant to cooperate. They had already had unsettling encounters with doctors who had suggested, among other Draconian measures, mass sterilization to prevent them from passing the disease on. Wexler promptly organized a town meeting. "We explained that it was a very long-range study," she says, "and that we were trying to find the cause of the disease. It might not help them, but could help their children and grandchildren."

Wexler told the villagers that her mother had died of the disease and that she herself might be stricken. She, too, had contributed a skin sample to be analyzed, she explained, and showed them the tiny biopsy scar on her right arm. "They really understood that," Wexler says, "and I think they soon realized that our intention was not to cause them harm. I became sort of like a family friend, with syringe."

As the pedigree chart grew, it became evident that virtually all of the villagers were probably descendants of a woman named Maria Concepcion Soto, who had lived in a village on the south shore of Lake Maracaibo in the early 1800s. Where she got the Huntington's gene is a mystery, but Wexler has a theory. "There was a lot of trade with Europe back then," she says. "Presumably some European sailor got down to Maracaibo and had a little dalliance, and a child, probably Maria, was born with the Huntington's gene."

Soon after the study began, Wexler realized that it would be prudent to photograph each villager involved, simply to keep track of relationships and make certain that each blood sample was matched with its actual owner. The Polaroid shots were pasted next to the appropriate square (male) or circle (female) on the rapidly expanding pedigree chart posted on the wall of the team's headquarters. "When the villagers came in to give blood," says Wexler, "we would say, 'Show me your relative.' And they would go and point, 'That's my great-uncle.' We would draw the pedigree from there and figure out how they were related."

The skin biopsies (which were found to be unnecessary and eventually discontinued) and blood samples collected by Wexler's team were sent to the Huntington's Center, a research facility at Massachusetts General Hospital. There, James Gusella, a Canadian scientist, was using molecular biology to probe the genetics of Huntington's disease.

In 1979, before samples from Wexler's Venezuelan family began arriving at the center, Gusella had decided that his best route was to launch a genetic-linkage study of a large Huntington's family. He turned to geneticist P. Michael Conneally at Indiana University, where a computer database of Huntington's families had been established. Searching through the Indiana collection of pedigrees, Gusella picked the largest—an Iowa family of 35 living members, 14 of them afflicted with Huntington's disease—and asked Conneally to collect and send him blood samples from each person.

To ensure an ample and steady supply of the white blood cells from which he would extract the DNA, Gusella infected the samples with Epstein-Barr virus so that the cells would multiply endlessly.

After culturing the infected cells until tens of millions of them were available, Gusella extracted their DNA, using an enzyme to digest the cellular protein and dissolving other cell components with organic solvents. His goal was to seek out RFLPs that were always inherited only by those family members afflicted with the disease. As suggested by David Botstein and his colleagues in their landmark 1980 paper, such markers must lie close to the disease gene on the human chromosome. "We anticipated looking at many pieces of DNA that showed variation until we hit one that was in the right location," Gusella recalls. "We didn't even know which chromosome to look on; it could have been anywhere."

But first Gusella had to determine that there was enough genetic variation in human beings to carry out the study. "At that point," he says, "one of the major arguments against this kind of study was that you wouldn't be able to find markers. Not just markers that were linked to the disease—you wouldn't be able to find markers at all."

Fortunately for Gusella and for those hunting other genes, it was not necessary (in fact, it was impossible) to seek out distinctive variations, or RFLP markers, by comparing lengths of the tiny DNA fragments under a microscope. A Scottish scientist, Edwin Southern, had already devised an ingenious and more manageable way to look for RFLPs: the Southern blot. To create a blot, Southern used a restriction enzyme to slice up a segment of DNA. The resulting fragments are placed at one end of a gelatinlike slab and exposed to an electric field with its positive polarity at the opposite end of the slab. Because DNA fragments carry a negative charge, they are pulled toward the other end, the smaller fragments moving more rapidly than the larger ones.

Eventually this process, gel electrophoresis, sorts out DNA fragments by size and strings them out in a column running down the gel.

Simultaneously, similar processes are occurring in adjoining lanes of the gel with DNA taken from the identical chromosomal region of other family members and fragmented by the same restriction enzyme. After the fragments are chemically denatured, a nitrocellulose membrane is placed over the gel to blot up the fragments, as a blotter absorbs ink. Still in their strung-out positions on the membrane, the fragments are washed over with a DNA probe, a sample of radioactively labeled, single-stranded DNA from the same chromosomal region. The radioactive DNA promptly seeks out and binds to the complementary sequences on the DNA fragments.

Finally, when an X-ray film is placed on the nitrocellulose membrane, the radioactively tagged fragments leave an image on the film that corresponds to their position on the membrane. Scanning the film, which displays patterned columns of dark bands, RFLP hunters can detect any differences in the fragment patterns between family members. If a difference exists between columns, say in the number or size of the fragments, a RFLP lies in the test segment.

To his relief, Gusella began finding those differences. By the summer of 1982, he had in the refrigerator a collection of DNA probes detecting enough variation to justify getting started. His group began analyzing different segments of the Iowa family genome, checking the X-ray film, comparing RFLPs with the pedigree of the family, and recording the resulting data. When they had accumulated the data for only the first 14 segments, they sent them to Michael Conneally, who had the software and computing power necessary for the proper analysis. "It isn't clear by eye that the correlation is there," Gusella explains. "The way you assess the level of co-inheritance is by computer analysis."

Luck was with Gusella. "You can hit it early, or you can hit it late," he says. "We hit it early." Conneally's computer showed that one of the first 14 DNA segments, sliced out of chromosome 4, contained a distinctive marker that correlated with the inheritance of the disorder. While that finding was highly suggestive, it still did not constitute absolute proof that the Huntington's gene lay somewhere close to that marker on chromosome 4.

The problem was that the 14 cases of Huntington's in the Iowa family were too few to prove the point. Conneally's computer put the odds at 35 to 1 in favor of the marker's being the one for Huntington's, but those odds were far too small to convince the scientists. "It was suggestive that we were going in the right direction," says Gusella, "but it wasn't proof. It could have happened by chance."

By 1983, blood samples from the huge Venezuelan family were arriving at Gusella's lab in ever greater volume. Analyzing the promising chromosome 4 segment of DNA from the first small group of the Venezuelan samples, Gusella again got a suggestive correlation between the same marker (which in these samples carried a pattern distinctive to the Venezuelan family) and inheritance of the disease. "The proof was still not strong enough," he says, "but by this time I was convinced that it was going to hold up." His group immediately turned to the remaining Venezuelan blood samples, which represented many victims of the disease, extracted the DNA, and analyzed it. In mid-July of 1983, they sent the data off to Indiana.

This time there could be no doubt. Conneally's computer, says Gusella, "clearly showed co-inheritance." It placed the odds at 1,000,000 to 1 that the marker lay near the Huntington's gene. Gusella, then only 30, published the results of his work in November 1983 in *Nature* magazine. It marked the first time that the recombinant DNA technique was used to find the location of a gene not previously assigned to a specific chromosome. "That basically is still the system that is used," Gusella says. "We demonstrated the success of the technique. And the actual proof that it could work is what convinced a lot of people to get into the field."

The discovery of the Huntington's markers led quickly to the development of a test that can determine with an accuracy of 96 percent if a person at risk is carrying the Huntington's gene. But that was hardly enough to satisfy Gusella, now an associate professor of genetics at Harvard Medical School. Since 1983 he, along with six other collaborating laboratories, has been steadily zeroing in on the actual gene, finding new markers even closer to it. "When we first found our marker," he says, "we learned that the Huntington's gene was somewhere on chromosome 4. So we had narrowed the search from three billion base pairs [the entire genome] to about 200 million [chromosome 4]. Then, by finding new and closer markers, we found that it was on the short arm of chromosome 4, and that narrowed it to about 70 million bases."

Another technique placed the gene even more precisely. Researchers were already aware that Wolf-Hirschhorn syndrome, characterized by mental retardation and a misshapen skull, resulted from a missing tip of the short arm of chromosome 4. By analyzing DNA from people with this syndrome, as well as that of normal people, Gusella discovered that unaffected people carried the marker for Huntington's, whereas Wolf-Hirschhorn victims did not. This proved that the marker lay somewhere on the tip, within 15 million base pairs from the end of chromosome 4's short arm.

By early 1991, researchers had narrowed their target area further, to a stretch of six million base pairs. They are now focusing on two promising segments within the six-million-base stretch, using the still-growing pedigree and the bank of cell cultures from Venezuela. One segment, at the tip of the chromosome, is about 100,000 base pairs long and has been completely cloned, which means that its sequence can be determined. The other segment, containing some two million base pairs, is located several million bases in from the tip and is in the process of being cloned. Now, Gusella says, "it's a matter of going in and looking at the genes that are there and trying to find one that is defective."

The reason he launched his hunt for the Huntington's gene, says Gusella, "was not so much to develop a presymptomatic test. It was to give us a handle for isolating the disease gene." When the gene is isolated, the accuracy of the Huntington's test should reach 100 percent because it will detect the presence of the actual gene rather than just the nearby markers.

The major benefit, Gusella says, is that "you will have a piece of DNA that encodes a protein that actually leads to the symptoms. You will therefore have the basis for determining how it does that. And how it does that is the knowledge that is essential if you want to find a therapy."

Continuing her role in the search for the gene, Nancy Wexler returns annually to Lake Maracaibo with a research team, steadily building the pedigree of the Venezuelan family and collecting more and more blood samples. Her enormous pedigree chart, posted on the wall outside her small office at Columbia University, now extends more than a hundred feet down the corridor and has grown to include more than 10,000 members of the family. She has taken blood samples from some 2,000 of them, and Gusella continues to analyze the DNA of the newest donors as he tracks the gene down to its precise lair on chromosome 4.

In fact, the very size of the Venezuelan pedigree makes it a valuable resource for tracking not only the Huntington's gene but other disease genes as well. Some of the families have other genetic disorders, and with the help of the huge pedigree and DNA analysis, medical researchers hope to trace those diseases to their genetic source. The same data can be used to help create the high-resolution linkage maps that will play such an important role in the genome project. "It's an extraordinary family," says Wexler admiringly, as she surveys her huge chart. "The pedigree is a big genetic playground. Whatever idea you have, you could probably test it there. It's a gold mine for genetic research and for the Human Genome Project."

HOT-WIRED

▼

For fast help with reimbursement for nonionic contrast media, call the Squibb Reimbursement Hotline at 1-800-842-4296.

ISO**VUE**® ...*enhances your image*
(iopamidol injection)

BRILLIANT FILMS

The Squibb Diagnostics case study series provides full-size duplicate films as diagnostic challenges in radiology. CME credits for this series are awarded by the Albert Einstein College of Medicine in New York.

SQUIBB™
Diagnostics
A Bristol-Myers Squibb Company

3/91

CHAPTER FIVE

WALKING, THEN JUMPING TOWARD GENES

In the years before the formal launch of the genome project, gene hunters, encouraged by James Gusella's success, embraced the RFLP strategy for tracking down long-sought genes. In 1985 they discovered three RFLPs that narrowed the search for the cystic fibrosis gene to chromosome 7. By the same means, the gene responsible for a rare form of familial colorectal cancer was traced to chromosome 5; the gene for retinoblastoma to chromosome 13; neurofibromatosis to chromosome 17; Duchenne muscular dystrophy to the X chromosome; and one type of familial Alzheimer's disease to chromosome 21.

The importance of finding these RFLP markers, writes Eric Lander, "is hard to overstate. In most cases, scientists have literally no clue about the fundamental cause of the disease. Biochemical studies frequently turn up a myriad of abnormalities, but they cannot distinguish cause from secondary effect." Finding a genetic marker, he says, confines the search "to a neighborhood a thousand times smaller" than the length of the entire genome.

But as Gusella discovered, finding the approximate location of these genes on a chromosome, tedious as it was, was merely the prelude to an even more daunting task: isolating the actual gene and sequencing it. Just how formidable that effort can be is best illustrated by the long, tortuous, but ultimately successful search for the cystic fibrosis gene.

Cystic fibrosis is the most common genetic disease among Caucasians in the U.S., and the search for the culprit gene, spurred in part by ample funding from the Cystic Fibrosis Foundation, was intense. In the

Dr. Francis Collins, a molecular geneticist at the University of Michigan, led one of the teams that helped to discover the location of a gene for cystic fibrosis on chromosome 7 in 1989.

many victims of this disease—one in 25 American whites is a carrier and as many as one in 2,500 is stricken—a thick layer of dehydrated mucus blocks air passages and plugs the ducts that transport vital enzymes from glands like the pancreas. Although CF was once fatal in childhood, better medical care has now prolonged the life of victims, enabling half of them to live beyond the age of 25.

By 1987, researchers had narrowed the search for the CF gene to a 1.7 million-base-pair stretch between two markers on the long arm of chromosome 7, and the race was on to isolate the actual gene. Among the contenders were a British team led by Robert Williamson, a geneticist at St. Mary's Hospital in London; another group headed by molecular geneticist Francis Collins at the University of Michigan in Ann Arbor; and a third team led by geneticist Lap-Chee Tsui of the Hospital for Sick Children in Toronto.

In April of that year, Williamson announced that his group had found a highly promising candidate for the gene, but further research proved that the newly discovered gene was also found in people unaffected by cystic fibrosis. The gene did serve as a closer marker for the

CF gene, however, reducing the region of search to about one million base pairs.

To move from a marker toward the target gene, researchers use a technique called walking. This tedious process involves starting from a known marker, slicing out a section of DNA a few thousand base pairs in length, cloning it in *E. coli*, then chopping the resulting clones into short segments with restriction enzymes. By sorting and analyzing these tiny fragments with the same gel electrophoresis technique that was employed by Gusella, scientists can determine which fragments have stretches of sequences in common (proving that they overlap) and then reassemble them in the proper order. Next, taking each fragment in turn, they search for markers or other evidence of the gene. If they find none in any fragments of the first section, they slice out the next section and repeat the process.

In an effort to quicken the pace, molecular biologists turned from cloning DNA in the *E. coli* plasmid, which can hold only small sections of DNA (and consequently confines genome walkers to tiny steps), and created a vehicle, or vector, that could accommodate large segments of DNA. They created this commodious vector, called a cosmid, by attaching DNA segments from a virus that infects bacteria to a 40,000-base-pair section of human DNA, producing a hybrid creation that could infect *E. coli*.

Like a real virus, the cosmid replicated inside the bacteria, producing ample supplies of 40,000-base-pair clones. Still, says Dr. Collins, "on average, you can cross only 20,000 base pairs a step. So to cover a million base pairs requires 50 such steps. Each step will take a month to six weeks." That means a minimum of 50 months, and Dr. Collins has found it difficult to interest graduate students, who do most of this tedious labor, in such projects. "It's even worse," he says, "since there are some regions of DNA that are unclonable. Sooner or later you could be sure that you'd get to one of those, and then you'd be stuck."

To avoid the pitfalls of chromosome walking, Dr. Collins and his group resorted to "jumping" their way across the suspect segment of DNA. Starting at a known RFLP and using restriction enzymes, they cut large 100,000-base-pair sections from the genome and used another kind of enzyme to curl these strands into closed circles. In each DNA circle, the two ends that join together are the DNA sequences that were initially farthest away from each other. "So now what you do," Dr. Collins explains, "is to clone just these junction pieces and throw out the rest." He and his group then searched the cloned pieces for a new RFLP, and DNA samples from a family with cystic fibrosis were analyzed to see if that marker consistently appears in victims of the disease.

When that finally happens, says Dr. Collins, "you've reached a region where there are absolutely no crossovers and there's a good chance that you're standing on top of the gene. That's when you've exhausted what you can do with genetic mapping."

In Toronto, meanwhile, Lap-Chee Tsui took 250 random samples of DNA from the suspect region of chromosome 7 and compared them with DNA from cystic fibrosis patients. Two of the samples tended to match the CF strands in families with the disease. At this point, the Toronto and University of Michigan teams decided to join forces because, as Dr. Collins explains, "our search parties, working together, could cover the territory a lot faster." By January 1989, they were down to a 300,000-base-pair region, and to avoid inadvertently jumping over the gene they were seeking, they went back to the time-consuming walking method, gradually zeroing in on one promising stretch.

Now the joint team made use of the appropriately named zoo blot. This technique is based on the fact that, as the animal kingdom evolved, genes essential to certain life processes were conserved and are found today, largely unchanged, in human beings. By comparing their DNA segment with DNA taken from chickens, cows, mice, and hamsters, the researchers found striking similarities between the human and bovine electrophoresis patterns. "There were several pieces of DNA that showed conservation," Dr. Collins says. This made it all the more likely that the Toronto team's segment of DNA contained the gene.

To determine whether the gene was the one responsible for cystic fibrosis, the researchers turned to a clone "library" that biochemist Jack Riordan, a member of the Toronto team, had painstakingly constructed over a two-year span. It consisted of hundreds of DNA samples copied from messenger RNA in the sweat glands of CF victims. Riordan used the following logic: cystic fibrosis is characterized by, among other irregularities, an excess of chloride in sweat. Because the disorder is probably associated with some faulty mechanism in the sweat glands, the cells of the sweat glands must contain an aberrant gene that expresses itself and is responsible for a flawed protein. The error in the genetic coding will therefore show up in the messenger RNA, and DNA copied from the RNA will also contain the coding flaw.

Comparing DNA samples from the sweat-gland library with their DNA segment, the Toronto and Michigan groups in the summer of 1989 finally found a gene that matched their DNA segment. It stretched across 250,000 base pairs of DNA and carried specifications for a protein made up of 1,480 amino acids (defective cystic fibrosis transmembrane regulator, or defective CFTR). And what was the flaw?

Comparing the CF gene with its normal counterpart, they found three base pairs missing, a genetic word consisting of the code letters CTT.

"What this does," says Dr. Collins, "is cause the loss of a single amino acid [phenylalanine]. And that makes the difference." To make certain that this mutation was not simply a harmless variation in a string of code letters between genes, the team compared the chromosome 7 DNA from dozens of CF victims with the equivalent DNA from healthy people. "In the initial study," Dr. Collins says, "we found that about 70 percent of CF chromosomes carry this mutation and that none of the normal chromosomes do. This is the mutation, and this is the gene." At that point in 1989, Dr. Collins was hopeful that only a small number of other mutations were responsible for the remaining 30 percent of cystic fibrosis cases and that a single test could soon be developed for all carriers of the disease gene.

But in 1990, Tsui reported that several laboratories probing the culprit gene in victims of the disease had already found an additional 20 or more rare mutations. These findings suggested the existence of even more mutations, each occurring in a different patient, and dimmed the prospects for a universally accurate CF test. Still, a test that can identify carriers of the CF gene who have the mutation found by Tsui, Riordan, and Dr. Collins (now estimated to be 75 percent of all carriers) is already available. But because it is not all-inclusive, it is being recommended only for people with a family history of the disorder.

The CF story continued to develop with startling speed as 1990 drew to a close. In separate experiments, two research teams isolated cells from CF patients and then infected them with viral vectors carrying copies of the healthy CF gene. With the good gene in place, these cells functioned normally, giving researchers hope that cystic fibrosis could eventually be cured by gene therapy.

CHAPTER SIX

A BIG PROJECT FOR A SMALL SCIENCE

T hough the greatest human endeavors have been built on the genius, inspiration, and drive of many people, the basic concept for such endeavors often arises in the mind of one person. In the case of the Human Genome Project, that individual probably was biologist Robert Sinsheimer.

The process that led Sinsheimer to his inspired vision was rather circuitous. In his role as chancellor of the University of California at Santa Cruz in 1984, Sinsheimer felt that his greatest challenge was to enhance the reputation of the school, one of the newer and lesser-known of the university's nine campuses. A quick way to accomplish that goal, he thought, was to establish the Santa Cruz campus as the base for a large and prestigious project. But what kind of project?

Sinsheimer was well aware that "in areas like physics and astronomy, scientists were not hesitant to propose multimillion-dollar projects if they felt they were needed and important and desirable." Biology was another story. Sinsheimer knew that his discipline was a "small science," its experiments and projects generally carried out by a handful of scientists using relatively inexpensive equipment. Still, he wondered, were there scientific opportunities in biology that were being overlooked "simply because we were not thinking on an adequate scale"?

Earlier, at Iowa State University and then at the California Institute of Technology, Sinsheimer had distinguished himself by purifying Phi X174, a bacterial virus, genetically mapping it and demonstrating that it consisted of a single-stranded ring of DNA. His work prompted the

British scientist Frederick Sanger to apply his newly developed sequencing technique to Phi X174. By 1977 Sanger had identified and placed in order all 5,400 chemical code letters in its DNA. This marked the first time that the genome of any organism had been completely sequenced. It also made a lasting impression on Sinsheimer.

By 1984 other scientists had sequenced entire viral genomes as large as 50,000 base pairs, and work was under way on the genome of the *E. coli* bacterium, which contains 4.7 million base pairs. That same year, the possibility arose that $36 million, a gift to the University of California toward the construction of a large telescope, could be diverted to another project.

An idea suddenly took shape in Sinsheimer's mind: what about the human genome, with its estimated three billion base pairs? Only a tiny fraction of 1 percent of those bases had been sequenced. Sequencing the entire human genome, which would require a huge biological project, could bring enormous benefits to medical science and biology. And if that project were based at Santa Cruz, it could put his institution on the map.

Sinsheimer wasted no time. In November 1984, in a letter to the president of the University of California, he proposed using the $36 million gift to establish on the Santa Cruz campus an Institute to Sequence the Human Genome. Soon afterward, however, negotiations over diverting the funds broke down, and the university reluctantly returned the money to the donor.

But Sinsheimer pressed on. "If we really knew the sequence of the human genome," he says, "we would have the blueprint for how human beings are made. I had done some crude calculations in my head that suggested it was obviously nothing you could do immediately, but it was not an inconceivable goal." The cost, he figured, would be about $25 million a year, an estimate that later proved to be woefully short of the mark. To test his assumptions Sinsheimer invited leading biologists, theorists, experimenters, and computer experts to a workshop at Santa Cruz in May 1985. Among those attending were David Botstein, then at MIT; Nobel laureate Walter Gilbert, the co-inventor of a chemical technique for sequencing DNA; and Dr. Leroy Hood, the Caltech biologist who had pioneered automation of the sequencing process.

At the start of the workshop, Sinsheimer's idea was generally greeted with skepticism; there were doubts about the practicality of such a huge biological effort. Dr. Hood, now an enthusiastic participant in the Human Genome Project, concedes that he had misgivings. "I thought Bob Sinsheimer was crazy," he says. "It seemed to me to be a very big science project with marginal value to the scientific community. But the more I thought about it, the more I saw that very important benefits—in terms of both basic science and medical science—would accrue."

By the time the meeting ended, Sinsheimer says, "most people were convinced that it was a feasible project, but I'm not sure the consensus was that it ought to be done." Undaunted, he wrote a summary of the workshop, including paragraphs that seem highly prescient today on such subjects as physical and genetic mapping, sequencing, computer requirements, and potential difficulties. "As events have proven," he says, "it was an idea whose time had come." To hasten the arrival of that time, Sinsheimer distributed the summary to all the workshop participants and to other prominent scientists in the field.

Charles DeLisi, the director of the Department of Energy's Office of Health and Environmental Research, had also been pondering an assault on the genome. The Energy Department and its predecessor, the Atomic Energy Commission, had long been involved in a biological program to study the effects of radiation on humans and their DNA. They had assembled a large team of skilled molecular biologists, designed elaborate devices for separating out and sequencing human chromosomes, and amassed an extensive library of human DNA samples, capacious data banks for storing genetic information, and big computers for analyzing it.

But, says Sinsheimer, "they'd pretty much learned what they were going to learn about radiation effects. So in some ways it was a well-developed program in search of a mission." DeLisi stepped into the breach. Early in 1986, he convened 50 experts from around the world in Santa Fe, New Mexico, to consider the prospects of sequencing the human genome by the turn of the century. This time virtually all the participants shared the sentiments of Walter Gilbert, who declared, "The total human sequence is the Holy Grail of human genetics. It would be an incomparable tool for the investigation of every aspect of human function." Rather than having the sequence revealed by bits and pieces as scientists investigated DNA segments of particular interest, Gilbert favored starting immediately on an effort to sequence the entire genome. "Why not do it once and for all on an industrial scale?" he asked.

Added impetus came in March 1986 when Nobel laureate Renato Dulbecco, writing in *Science* magazine, urged the launch of a genome project, citing the benefits it could offer in solving the mysteries of human cancer.

Encouraged by the groundswell, the Department of Energy acted quickly, announcing plans for a human genome initiative that had as its ultimate goal the complete sequencing of human DNA. Word of the

ENHANCING PARTNER

▼

Squibb Diagnostics wants to be your partner in enhancing the quality of diagnostic medicine.

ISOVUE® *…enhances your image*
(iopamidol injection)

EDUCATIONAL TELEVISION

▼

Squibb Diagnostics produces in-service films for radiologists, radiology departments and administrators, including *The History and Development of Contrast Media* and *Conversion to Nonionic Contrast Media.*

SQUIBB™
Diagnostics

A Bristol-Myers Squibb Company

Energy Department's decision caused consternation in the scientific community, which was still sharply divided over the prospects and the goals of so huge an endeavor.

David Baltimore, a Nobel laureate biologist and then the director of MIT's Whitehead Institute, voiced the concern of many scientists. He feared that so massive a project would have much the same impact on biology that the shuttle had had on U.S. space science—siphoning grant money and talent away from other smaller and far less costly endeavors that had proved so productive in the past.

It was the pell-mell rush to sequence the genome that bothered David Botstein. "Knowing the sequence," he declared, "is like having the writing on the pyramids without having the Rosetta stone. It means nothing. It could be a laundry list or the last will and testament of some emperor. What you need is the means to understand the sequence, and the means is biological experimentation, most of which will not be done on humans, for obvious reasons, but on other organisms."

But opponents of the sequencing project realized that they were fighting a losing battle. "The idea is gaining momentum," said Baltimore. "I shiver at the thought."

The Energy Department adamantly moved ahead. In the fall of 1987, it designated its Lawrence Berkeley and Los Alamos national laboratories as genome research centers, allocated funds, and began work on what was quickly shaping up to be a biological megaproject. Soon afterward, the Lawrence Livermore Laboratory was added.

DeLisi had clearly seized the initiative. His bold action may have served as a wake-up call to dubious and hesitant biologists outside the Energy Department. In February 1988, after 14 months of study, a National Research Council committee composed largely of university-based scientists (and including such luminaries as Watson, Botstein, and Dr. Hood) proposed a genome project that seemed to satisfy almost everyone, including Baltimore.

In a 102-page report entitled "Mapping and Sequencing the Human Genome," the committee recommended a 15-year program that would aim first at creating genetic (linkage) and physical maps of the human genome. With ever-increasing accuracy, the maps would help to identify and locate genes on the individual chromosomes. At the same time, work would proceed on improving the technology and lowering the cost of sequencing until, during the latter part of the 15-year project, the entire three billion base pairs of the genome would be sequenced.

The committee proposed scaling up quickly to $200 million annually in federal funds to support the project. But, addressing the concerns of many biologists, it emphasized that "these additional funds should not be diverted from the current federal research budget for biomedical sciences." It also recommended that a single federal institution be designated to oversee the project and provide funding to research groups chosen through peer review. "Acquiring a map, a sequence, and an increased understanding of the human genome," the committee concluded, "merits a special effort that should be funded specifically for this purpose. Such a special effort in the next two decades will greatly enhance progress in human biology and medicine."

An impressive array of scientists gave eloquent and enthusiastic testimony at a congressional hearing convened in April 1988 to consider funding the genome project. Among them was James Watson. "I see an extraordinary potential for human betterment ahead of us," he declared. "We can have at our disposal the ultimate tool for understanding ourselves at the molecular level. . . . The time to act is now."

Congress rose to the challenge. For fiscal 1989, it allocated $46 million for genome research. By awarding $28 million of that total to the National Institutes of Health, Congress made the NIH the de facto lead agency in the project. But the Energy Department was hardly ignored. It received the remaining $18 million.

Warming to the task, the NIH in the fall of 1988 named James Watson as project director, an appointment that was universally acclaimed. Even the Energy Department scientists were impressed. "It's an incredible appointment," said Robert Moyzis, director of the department's Center for Human Genome Studies. "First Watson was personally responsible for figuring out the actual structure of DNA, and now he's at the point where he's likely to see the entire sequence of the human genome uncovered, all during his lifetime." David Botstein, now at Stanford University, was equally enthusiastic. "Jim Watson is as effective an advocate for good science as anyone could be. He has the ability to recruit other good people to help him. It's an appointment that's almost ideal."

While Watson, with the help of his Genome Advisory Committee, faced a formidable task in organizing the ambitious project, he did not have to start from scratch. Long before the concept of a genome project arose, scientists around the world were independently probing isolated bits and pieces of the genome, searching for disease genes, devising automated machinery to sequence DNA, establishing repositories for actual DNA samples, and setting up data banks to house the enormous amount of information they were accumulating.

By the time Watson assumed his new post, some 4,500 of the 50,000 to 100,000 human genes had been identified, and about 1,500 of them

Dr. Victor McKusick is the editor of *Mendelian Inheritance in Man*, the authoritative and continually updated compilation of all the known information about all the verified genes—more than 5,000 so far.

had been roughly located on the chromosomes. Segments containing several million of the genome's three billion code letters had been sequenced, and a number of centers were independently processing and cataloging the flood of data resulting from these efforts.

Dr. Victor McKusick, for one, has been cataloging genes since 1959. Supported by grants from the Howard Hughes Medical Institute, the Johns Hopkins geneticist compiles all the known information about each of the verified genes in his voluminous and regularly updated publication, *Mendelian Inheritance in Man*. In 1987 he introduced an on-line version of the catalog that is accessible by computer to scientists around the world. As of March 1991, it contained information about some 5,200 genes. "That's an impressive figure," Dr. McKusick says, "but we still have a long way to go."

The peripatetic Dr. McKusick was also the founder and first president of the Human Genome Organization (wryly referred to as "Victor's HUGO"), a group formed in 1988 in Montreux, Switzerland, by 42 scientists representing 17 nations. Described as "the UN of gene mapping," HUGO is now headed by Dr. James Wyngaarden, the former director of the NIH. The group is establishing three data-collection and data-distribution sites, one each in Japan, North America, and Europe, and will also open an office in Moscow.

By far the largest repositories of genetic information are the DNA Data Libraries (DDL), established in 1980 by the European Molecular Biology Laboratory in Heidelberg, Germany, and the Los Alamos National Laboratory's GenBank. The DDL now contains about 33 million base pairs of sequence from the DNA of humans and several hundred other organisms. Collaboration with the DNA Data Bank of Japan is expected to add another 10 million base pairs of sequence to the collection. Other, more specialized data banks already contain vast stores of genetic data.

Still, the data storage, software, and computer capabilities of all these institutions combined pale in comparison with the eventual needs of the genome project. In addition to storing and interpreting the three billion code letters in the human genome, the data centers will be required to handle the billions more from the genomes of other organisms. Doctors and researchers tapping into the database network by computer will want individual genes identified by sequence, by the traits they control, and by the diseases they cause. And supercomputers using highly sophisticated software will be needed to search the sequences between the genes for meaningful information.

"Even if we had the sequence today," says Charles Cantor, former head and now principal scientist of the Energy Department's genome efforts, "we couldn't use it properly. We don't have the knowledge or the computer smarts to interpret that information."

To develop those smarts, the genome project has established a joint NIH-Energy Department Informatics Task Force. One task force member, Tom Marr, a computer specialist and biologist at Cold Spring Harbor Laboratory, defines his group's greatest challenge with a question: "How does one construct an algorithm to pick out biologically significant DNA?" Programmed with the proper algorithm (a step-by-step list of instructions specifying how information will be handled to solve a particular problem), a computer could quickly distinguish the biologically important exons from the inert introns in a gene. It could determine where one gene ends and another begins, or how a collection of overlapping DNA segments sliced from a chromosome can be reassembled in the proper order to form a physical map of that chromosome. Endowed with these capabilities, computers could greatly accelerate both the mapping and sequencing of the human genome, while significantly reducing costs.

How to attack these and other data problems and how to coordinate the efforts of dozens of independent genetic research centers are

PATIENT-WISE

▼

Through easily understood educational videos and brochures produced by Squibb Diagnostics, many patients' fears about upcoming imaging procedures are allayed.

ISOVUE® ...*enhances your image*
(iopamidol injection)

1

READING GLASSES

▼

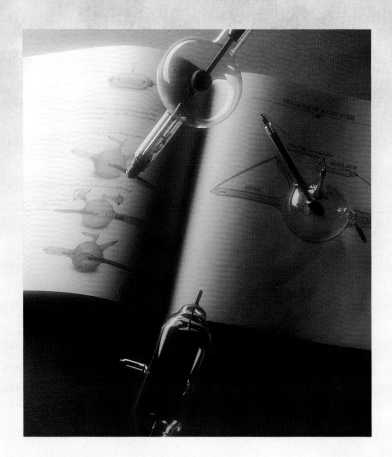

Squibb Diagnostics has recently published a limited edition of Dr. William Shehadi's *Reflections on the Radiology Years.* This new book serves as a guide to Dr. Shehadi's vast collection of glass-bulb x-ray tubes, now on display at the Johns Hopkins University School of Medicine.

SQUIBB™
Diagnostics

questions still being resolved by Watson and his advisory committee as they expand their organization. But considerable progress has already been made. The NIH and DOE groups are coordinating their efforts and have drawn up a rough timetable for achieving goals within the 15-year life of the project.

Watson set October 1990 as the time when the 15-year clock would begin ticking. "That was a good time to start," explains Norton Zinder, the Rockefeller University biologist who heads the advisory committee, "because that was the start of the government's fiscal year, and everything is always contingent on how much money you get. If we don't get close to the $200 million annually after a few years, then of course all bets are off." Supporters hope Congress will continue to move in the right direction. For fiscal 1990, which began in October 1989, it raised the ante, appropriating a total of $86 million for the project, $60 million to the NIH and $26 million to the Energy Department. For fiscal 1991, $134 million was appropriated.

Given the necessary funding, the scientists plan during the first five years to create a high-resolution linkage map of the entire genome and a physical map that will locate segments of DNA along large stretches of the genome. Scientists will also sequence another 10 million or so of the human genome's three billion code letters, and some 20 million base pairs from the DNA of other organisms. In the process, they hope to improve the efficiency of sequencing technology by a factor of from 5 to 10. During the next five years, if all goes well, the genome will be completely mapped, a good portion of the 100,000 genes identified and located, and sequencing efficiency improved. Finally, by 2005, the project leaders hope that all the human genes will be identified and located and the genome completely sequenced.

Actually, Watson confesses, he would like to see the project completed by 2003. "It's a personal thing," he says. "By then, I will be only 75 and presumably still capable of appreciating it."

Watson and Cantor set up a joint committee to ensure cooperation between NIH and the Energy Department and to keep wasteful overlapping at a minimum. Energy Department scientists, for example, have already staked out four or five chromosomes and have begun mapping them. Watson's group will concentrate its mapping efforts on the remaining chromosomes.

In fact, says Norton Zinder, "the chromosomes will be the unit of management in this business. That's how things will be coordinated. People who work on the same chromosome will have workshops together and will share information, materials, and databases." One such gathering occurred in May 1990, when James Gusella, Ray White, and other leading gene mappers met at the Cold Spring Harbor Laboratory and agreed on group leaders who will provide a 200- to 300-marker reference map covering all of the chromosomes.

Some of the money awarded to NIH for the project will be spent on establishing several research centers, each with specific objectives. And large sums will be designated for worthy research projects. It will not suffice for scientists seeking those grants to propose, for example, simply studying a virus. "We're not interested in that," says Zinder. "Instead, we'll ask, 'How many nucleotides are you going to sequence? How many genes are you going to map?' That's what we want to know. This is applied work, not basic research."

As an important phase of the genome project, grants will also be made for studies of genes in other organisms, such as mice and fruit flies. "We've got to build a few places that are very strong in mouse genetics," Watson says, "because in order to interpret the human, we need to have a parallel in the mouse." Humans, mice, and lower organisms share many sequences of genetic code letters that are virtually identical. "Experimentation with lower organisms will illuminate the meaning of the sequence in human beings," says David Botstein. For example, genes that control growth and development in the fruit fly are virtually identical to oncogenes.

Investigation of lower organisms should have another payoff. Genome project directors have already agreed to provide half the funds for a joint U.S.-British effort to sequence the genome of the nematode worm *C. elegans*. This will be no small task. The *C. elegans* genome has some 100 million base pairs, and the largest genome completely sequenced by late 1990 was that of a herpes virus, cytomegalus, with 250,000 base pairs. Researchers hope initially to sequence a million base pairs a year, then scale up their efforts and complete the entire sequence in about 10 years. The object, says Zinder, is not only to learn what the worm's genome sequence is, but to develop techniques for faster and more efficient sequencing that eventually will be applied to the human genome. "*C. elegans*," he says, "will be our workbench."

Despite the difficulties and complexities of the task that lies ahead, Watson cannot contain his enthusiasm for the project that he directs. "Thirty years ago," he says, "my dream was that we'd know the structure of a virus. The fact that we've progressed so far that we might know the precise structure of a human seems to me to be wonderful. DNA is what gives us our genetic instructions. Of course I want to know what a mouse is, what a fly is, what a bacterium is. But deep down, I want to know what I am."

CHAPTER SEVEN

MAPPING, YACS, AND THE ULTIMATE GOAL

We started the project based on the assumption that we would be able to improve the technology tremendously. That guess is going to turn out to be true.

—Charles Cantor

The optimism reflected in Charles Cantor's pronouncement is not uncommon among the principals in the Human Genome Project. Given the proper funding, they say, the goals of the mammoth undertaking can be met, or even surpassed. But if the travails of Dr. Francis Collins and his cystic fibrosis gene hunters are any indication, the prospect for finding and identifying the remaining tens of thousands of unknown human genes would seem rather uncertain indeed. In fact, Dr. Collins says, "in many ways, cystic fibrosis is a very attractive and relatively easy target because of the high frequency of the disease." Large numbers of family pedigrees and DNA samples accompanying each, he says, "allowed us to really push the genetics to the point of narrowing down the candidate area to 500,000 base pairs."

For many less common but serious hereditary disorders, the pedigrees and DNA samples are scanty, and identifying the errant genes with the genetic tools now available would be even more difficult, if not impossible. For those reasons, Dr. Collins believes, "having the genome project completed all the way from the fine-resolution linkage map down to the sequence will be enormously helpful. I would say that for many diseases of major biomedical importance, the only way the genes will be found is if the genome project provides these kinds of aids."

The Human Genome Project is already financing linkage mapping at the University of Utah in Salt Lake City. There, under Ray White, a group of scientists has for several years been compiling a genetic linkage map of normal human chromosomes, drawing on a precious local resource—large, multigenerational Mormon families.

White, who began his linkage mapping at the University of Massachusetts Medical School, moved to Salt Lake City in 1980 because, as he says, "it very quickly became apparent that if one were going to make a serious commitment to this kind of work, Utah was the right place to do it. The Mormons still have lots of kids—many families have five, six, or more children, even today. When you're doing genetic research, the children are where the information is." The families, White says, "are just fantastic from a human geneticist's point of view. They go back six or seven generations to 1850, when the first founding population came in, and they keep exquisite family records." Also important to White's work, he says, is that the Mormons "still believe that science and medicine are basically good for the world, and they want to help."

Working with blood samples donated by Mormon families, White's group reported in 1987 that it had identified 475 RFLPs, or markers, on the human genome, each about 10 million base pairs apart. That map, and a similar one developed by the Collaborative Research Company in Bedford, Massachusetts, have already helped scientists zero in on several disease genes. White himself used his map to locate an oncogene that served as a close marker in the hunt for the cystic fibrosis gene. Still, he is not satisfied. His group is currently developing a RFLP map with much higher resolution, the markers perhaps only a few million base pairs apart.

When both low-resolution and fine-scale linkage maps are complete, White says, "workers trying to locate a disease gene will be able to choose and test a set of markers spaced at equal, large intervals along the chromosomes." Once they discover a linkage that narrows the gene's location to a specific chromosomal segment, he says, they can test markers from a fine-scale map of the region to find the tight linkage, just as Dr. Collins and Lap-Chee Tsui did when they narrowed their search to a 500,000-base-pair segment. From that point, a variety of molecular studies, including several used by Dr. Collins and Tsui, could be used to finally pinpoint and identify the gene.

But linkage maps, however helpful, do have their limitations. Though they indicate the proper order of markers along the genome, they do not reveal the actual physical distance between them. Instead they measure the chance, given as a percentage, that the markers will be separated during recombination, or crossover. A linkage map, explains David Botstein, "exists in genetic space, as measured by recombination, rather than in physical space, as measured in base pairs."

If the chance of recombination between two markers is only 1 percent—a figure determined by DNA analysis of family pedigrees—scientists say that the markers are a "centimorgan" (after the pioneering Thomas Hunt Morgan) apart. A 10 percent chance that they will cross over places them 10 centimorgans apart. On average, a centimorgan represents an actual distance of a million base pairs, but that measure is far from precise. Along some "hot" stretches of the genes, for example, recombination occurs frequently; on others, it rarely takes place. Consequently, a centimorgan in one region of a chromosome might represent only 400,000 base pairs, in another segment, perhaps two million.

These genetic mapping techniques produce a broad outline of the genome, identifying the relative positions of genetic markers on each chromosome. But for a more precise placement of genes a physical map is needed, and constructing one is a major goal of the Human Genome Project. Actually, one kind of physical map, assembled largely by Dr. Victor McKusick, already exists. It consists of a stylized image of each chromosome, showing its distinctive dark and light banding pattern (brought out by cytogenetic staining) and the approximate location of each known gene on that pattern. But the resolution is quite low; each band is hundreds of thousands of base pairs wide, and the gene could lie anywhere within a band.

A more accurate physical map, which genome project scientists are now beginning to assemble, will be composed of overlapping segments of DNA, precisely ordered and covering the entire length of the genome. It will enable scientists studying a segment of DNA to determine exactly where it lies on the genome. In the search for the cystic fibrosis gene, says Dr. Collins, "having overlapping clones of the region on chromosome 7 would have saved us two years of work."

Here again, the genome project will not have to start from scratch. Energy Department scientists have already established a clone library of all the human chromosomes. Using restriction enzymes, they have snipped each chromosome into pieces and cloned the pieces, each thousands of base pairs long, in colonies of *E. coli*. "Right now," says Robert Moyzis, of Los Alamos, "if you know a gene is on chromosome 4, rather than fishing around in the whole genome, you can go to the chromosome 4 library." Though the library saves time, however, it has yet to be cataloged, says Moyzis: "It's as if somebody took all the books in a regular library and threw them in a big pile on the floor."

Placing the tiny clones in order along chromosome 4 (let alone the

X

Duchenne muscular dystrophy

Becker muscular dystrophy

Chronic granulomatous disease

Retinitis pigmentosa-3

Retinitis pigmentosa-2

Agammaglobulinemia

Hemophilia B

Fragile X syndrome

Hemophilia A

Manic-depressive illness, X-linked

Colorblindness (several forms)

Diabetes insipidus, nephrogenic

The Human Genome

Y

XY gonadal dysgenesis

Using low-resolution mapping techniques, researchers have located many genes on the 46 human chromosomes. The chart at right shows the 22 autosomal chromosomes and the X and Y chromosomes with a partial listing of some of the genes that have been located so far. New sites are being found almost every month. The standard reference source for this catalog of the known human genes is Dr. Victor McKusick's *Mendelian Inheritance in Man*, which now contains the chromosomal locations of around 2,000 genes.

A human karyotype (right) is made by staining chromosomes at a stage of cell division when they are duplicated but not yet separated from each other. Such chromosome spreads are photographed under a microscope. The individual chromosomes are identified in the resulting photos, cut out, and assembled in order. The banding patterns on chromosomes are revealed by staining with fluorescent dyes. Each chromosome has a unique pattern of bands, with each band representing about 100 genes.

Many genes can be mapped to specific bands, thus indicating their general location on a particular chromosome, using methods like in situ hybridization. In the chart at right, genes are listed top to bottom relative to their position on the chromosome.

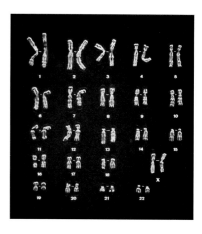

A HUMAN KARYOTYPE

Lift the page at right to expose the chart showing the human genome. Turn the chart over for an explanation of the levels of resolution needed to map gene locations on chromosomes.

entire genome) is a formidable task, but genome project scientists are facing up to it. "Physical mapping projects are akin to solving very difficult jigsaw puzzles," says Maynard Olson, a molecular geneticist at Washington University in St. Louis. "There are very large numbers of pieces all similar to one another, and some of the pieces are damaged in some fashion, so they just don't fit."

At the Human Genome Center at Lawrence Livermore National Laboratory, geneticist Anthony Carrano and his group are working on a physical map of chromosome 19. "The chromosome is 60 million base pairs long," he says, "and just to cover that span we need 1,500 cosmids [each 40,000 base pairs long] linked end to end." Because some cosmids are more clonable than others, the group will eventually work with about 8,000 cosmids, hoping to reassemble them in order.

Scientists at Livermore slice each 40,000-base-pair cosmid into about 70 pieces, all of which are tagged with blue, green, or yellow fluorescent dye. The size of each fragment (determined by running the fragments through the gel electrophoresis process, which separates them according to length) is then fed into a computer, along with the fragment lengths from other clones. The computer, Dr. Carrano says, has been programmed to ask how many fragments of the same size the clones have in common. "If they have a great number of fragments of the same size," he continues, "then we know they overlap."

To obtain even bigger jigsaw pieces, Maynard Olson used newly found restriction enzymes that cut sites occurring only rarely in the genome. This enabled Olson and graduate student David Burke to cut out segments of human DNA averaging 250,000 base pairs, and some as large as a million base pairs, in length. By splicing these giant molecules to segments of yeast DNA that control the replication of chromosomes, they created a hybrid concoction that could mimic a yeast chromosome. These yeast artificial chromosomes, or YACs, were then inserted into yeast cells. Perfectly at home with the disguised intruders, the yeast cells divided and replicated, duplicating the hybrids as well as their own chromosomes in every cell division. In a two-year project, Olson and his colleagues at Washington University produced a library of 60,000 YACs that represent the entire human genome.

Olson's creations have paved the way for what he calls "top down" physical mapping. Instead of piecing together tiny fragments of DNA in a "bottom up" mapping technique that eventually covers the whole genome, he suggests starting with each chromosome and moving immediately to YACs to produce a low-resolution physical map.

To place the giant YACs in proper order, Olson will sort them out by gel electrophoresis, which until a few years ago could handle only DNA segments that were less than about 50,000 base pairs long. In 1984, however, David Schwartz and Charles Cantor, then at Columbia, discovered that by rapidly pulsing and reversing the electrical field applied to the gel, they could sort out DNA segments as large as millions of base pairs in length.

That is the process Olson and others will use in producing a top-down map. Once a high-resolution linkage map becomes available, Olson says, scientists will have a landmark every few million base pairs or so. "If one finds all the YACs corresponding to those landmarks," he explains, "that would anchor the YACs into known positions. Then you could go about finding YACs near those you had anchored" until the entire genome has been filled in. This low-resolution physical map could be refined by breaking each YAC down into smaller cosmids and placing them in order.

In an effort to speed up the pace of this large-scale mapping, Olson has generously distributed complete sets of the laboriously produced YACs to six other laboratories. James Watson is impressed. "This is the way we hope others will act," he says.

The Energy Department's Moyzis likes Olson's approach. "The best strategy is to get the maps," he says. "If we can get a low-resolution physical map faster, then let's get that." At Los Alamos, he and Mary Kay McCormick have constructed more limited YAC libraries from individual chromosomes.

Though linkage mapping is well under way and physical mapping is making good headway, the ultimate physical map—a completely sequenced human genome—has barely been started. Of the three billion base pairs in the human genome, segments containing fewer than three million base pairs (less than a tenth of 1 percent) have been sequenced. By late 1990, the largest single segment of human DNA that had been sequenced was the growth-hormone gene, which is 150,000 base pairs in length.

The slow pace of sequencing reflects the complexity of each of the two ingenious but laborious techniques used in the process. One was devised by Walter Gilbert and Allan Maxam. It uses chemicals that specifically destroy either the A, T, C, or G nucleotides in radioactively tagged pieces of the DNA segment being sequenced, leaving fragments of DNA that terminate in a known nucleotide. The fragments are then run through the ubiquitous gel electrophoresis process.

In the method developed by Frederick Sanger of Cambridge University, who shared the 1980 Nobel Prize in Chemistry with Walter Gilbert, the DNA to be sequenced is separated into two strands, and enzymes are used to begin the growth of a radioactively tagged copy of one of the strands. The growth is then halted by one of four "chain terminators," each of which stops growth at either an A, T, C, or G. As in the Gilbert-Maxam method, this process produces DNA segments that terminate in a known nucleotide. Again, the pieces are sorted out by gel electrophoresis.

Both the Gilbert-Maxam and the Sanger methods, introduced in 1977, greatly increased productivity, accelerating the rate of sequencing from a frustratingly slow 10 to 100 base pairs that one person could achieve in a year to around 5,000 base pairs. Since then refinements and automation have increased the rate to as much as 100,000 bases per person-year. But as remarkable as the progress in sequencing has been, the current state of the art is woefully inadequate for the task ahead.

Addressing the problem in its 1988 report, the National Research Council's committee estimated that with current technology, the task of sequencing the entire three-billion-nucleotide sequence in the human genome "would require 30,000 person-years of work at a cost of $3 billion." That is clearly out of the question. Moreover, many scientists argue that to ensure accuracy, the effort would have to be duplicated at least once. And that is not all. To interpret the human sequence better, researchers would also need the genomic sequences of several lower life forms—the *E. coli* bacterium, the *Drosophila* fruit fly, the nematode worm *C. elegans*, and the laboratory mouse, among others.

The answer, says Dr. Leroy Hood, is more automation. "With the technology we have now," he says, "if we press, we can do 16,000 nucleotides a day. But before we can seriously take on the genome initiative, we will want to do 100,000 to a million a day. I can see technologies that put us at the 100,000-a-day level within five years, but if we work hard and think hard, we can probably do a lot better."

Current techniques of sequencing DNA are so labor-intensive that the cost can run as high as several dollars per nucleotide. If the genome project is to meet its goals, Dr. Hood says, that cost must come down. He believes that "a penny a nucleotide" is a realistic and attainable 10-year goal.

Dr. Hood's laboratory is experimenting with several different alternatives to today's technology. One involves using fluorescent labels rather than the radioactive ones now generally employed in gel electrophoresis. Current sequencing techniques can track only one reaction in each column, or lane, of the gel. But using green, blue, orange, and red fluorescent dyes makes it possible for scientists to speed up sequencing by running and keeping track of four reactions per lane.

Two other recent developments are already speeding the work of the genome project. The first, known as the polymerase chain reaction

(PCR), produces an ample quantity of specific DNA segments as long as a few thousand base pairs without the aid of host organisms such as yeast or *E. coli*. In an automated process conceived by Kary Mullis and developed by Henry Erlich and colleagues, all at Cetus Corporation, the DNA to be copied is denatured and a short DNA "primer" is annealed to each of the two separated strands. Exposed to the enzyme DNA polymerase, each primer extends into a new DNA strand and with the original strand forms the familiar double helix. Where there was one DNA segment there are now two.

The process is repeated again and again, the new copies separating from the original strands and both sets serving as templates for another amplification. Within three hours, more than a million copies of the original DNA segment have been produced, enough for most experiments. The development of PCR and automated sequencing machinery, says Charles Cantor, represents "a revolution in our ability to go from a raw biological sample—a cell, a hair follicle, sperm—to DNA sequencing." To speed up their mapping efforts, scientists are now seeking ways to extend the PCR technique to DNA segments hundreds of thousands of base pairs long.

PCR will also play an important role in an ingenious technique suggested by Olson, Dr. Hood, Cantor, and Botstein and promptly adopted by the genome project—sequence-tagged sites (STSs). Under this plan, researchers would sequence a segment of a few hundred base pairs (the STS) in, say, a 40,000-base-pair DNA clone under investigation, and would further identify two 20-base-pair sequences, one at either end of their STS, as primers. Then, in conveying information about their clone to a central data bank, they would include the sequence of their STS and its primers.

The process of tagging a clone with a distinctive sequence will be of enormous benefit to other scientists who want to work with it. Instead of asking the original investigator to send them copies of the clone, scientists will merely need to link up their computers with the data bank, obtain the primer sequences, and synthesize them. The primer sequences alone could be used with PCR to seek out their counterparts on a chromosome, thus locating the STS sequence and the DNA segment identical to the one used by the original investigator.

"I see STS as a way of distributing information," says James Watson. "You don't have to wait six months to get the clone." Also, by requesting that everyone use the STS labels for the segment of the genome they are studying, the project directors have instituted a common language, ensuring that when genetic and physical mappers describe a marker on a chromosome they are all referring to precisely the same location.

BRILLIANT FILMS

The Squibb Diagnostics case study
series provides full-size duplicate
films as diagnostic challenges in
radiology. CME credits for this series
are awarded by the Albert Einstein
College of Medicine in New York.

ISOVUE® ...*enhances your image*
(iopamidol injection)

'91

HOT-WIRED

▼

For fast help with reimbursement
for nonionic contrast media, call
the Squibb Reimbursement Hotline at 1-800-842-4296.

Perhaps most important, the STS approach eliminates a burden that project planners once considered necessary for success: a massive central repository where the hundreds of thousands of clones needed to cover the entire genome would be permanently stored for future sequencing. "Now," says Dr. Hood, "the repository is simply a computer with the PCR primer sequence for the STS."

The NIH's Genome Advisory Committee has proposed as the first major goal of the genome project a sequence-tagged site map of the genome with STSs spaced at 100,000-base-pair intervals along each chromosome. For the entire human genome, that would require some 30,000 mapped STSs. And to ensure that all of the genome researchers can communicate their findings in a universal language, the committee has recommended that existing sets of mapped DNA segments be converted to STSs.

More remarkable advances are in the offing. For example, Energy Department scientists, working with the recently developed scanning tunneling microscope (STM), announced that they had produced images of the DNA molecule. The images have such high resolution that the major and minor grooves between the coiled strands of the double helix are clearly visible. That startling view of DNA suggests that it may someday be possible to read the DNA sequences visually. Attainment of that goal would dramatically shorten the time needed for genetic diagnosis and bring closer the day when the entire message of life can be translated and read.

CHAPTER EIGHT

A MIXED BLESSING?

Know then thyself . . . the glory, jest, and riddle of the world.
—Alexander Pope

In laboratories from Cambridge to Ann Arbor, from Baltimore to Tokyo, Pope's wise counsel is being heeded. For all the hopes and promises of the genome project, however, the knowledge that it brings could prove to be a mixed blessing. Without proper safeguards, knowing one's own genome or having others know it can have dire effects both on individuals and on society.

The possibility that the information gleaned from their vast undertaking will be misused has not been ignored by scientists. As they isolate one disease gene after another, it becomes ever more apparent that their discoveries will pose a host of ethical dilemmas that society is not yet prepared to resolve. Their concern was expressed in a 1988 National Research Council report that was instrumental in gaining widespread approval for the Human Genome Project. "Without careful interpretation," the report warned, "information that links particular genes with disease can have harmful consequences for the people who carry those genes, quite apart from the disease itself."

Those consequences are manifold, involving psychological damage, social stigma, financial loss, and job and insurance discrimination. Without judicious policies and legal safeguards, the new genetic knowledge could tear at the fabric of society, exacerbating tensions over such issues as abortion, individual rights, privacy, and the role of government. Ray Gesteland, co-chairman of the human genetics department at the University of Utah, expresses the concern of people both in and outside of the Human Genome Project: "The technology is advancing at an incredible rate compared to our willingness to address the financial, social, and ethical questions this raises."

James Watson makes it a point to stress the importance of ethics at every meeting of genome project scientists. "In helping to lead this program," he says, "I want to emphasize that as we collect this knowledge the public should be protected. Who should know about our genes? Should they be public knowledge? My feeling is absolutely not. My DNA is my own private right to know. The object should not be to get genetic information per se, but to improve life through genetic information."

Watson's concern for ethics is not mere window dressing designed to divert opposition to the genome project. To help other project scientists, doctors, and the public with such issues, he established a working group as part of the Genome Advisory Committee and appointed Nancy Wexler to head it. In December 1989 he announced plans to spend as much as 3 percent of the project budget on grants for studies of ethical, legal, and social questions and on a series of nationwide town meetings to discuss such concerns.

The advisory committee will have a good base to build on. Many potential problems were anticipated by a presidential commission on ethics that considered the impact of genetic engineering, screening, and counseling and issued two reports on its findings in the early 1980s. Ethicists will also benefit from earlier experience with the psychological and social problems stemming from the development of tests for sickle cell anemia and Tay-Sachs disease, and for AIDS.

"The big difference that the genome project will make," says Wexler, "is in the multifold increase in the amount of information and the number of people affected. The difference in scale has absolutely monumental implications." Wexler cites the rapid development of the first CF test as an example. "One in 25 people are carriers of the CF gene, and there may be so many people who want the test that there might even be some move toward population screening."

The fact that tests are being developed for still-incurable diseases worries Arthur Caplan, director of the Center for Biomedical Ethics at the University of Minnesota. "If we don't get cures, if we only get some information about abnormalities, would that lead to more abortions?" he asks. "I don't know if having a 20 percent disposition to diabetes is a disease or an abnormality. I'm certainly not sure whether it morally justifies anyone aborting a fetus with that genetic profile. We haven't thought very much yet about how to draw that line between what is a disease and what isn't."

Utah senator Orrin Hatch strongly supports the genome project but recognizes that there are ethical considerations in any biomedical research. "As we advance with sequencing the human genome," he says, "we will have the ability to detect a genetic abnormality long before we have the ability for any kind of therapeutic intervention."

This concern is perhaps best illustrated by the discovery in 1983 of markers close to the gene for Huntington's disease. A positive test is tantamount to a death warrant for an otherwise apparently healthy adult; the gene is dominant, and its presence ensures the eventual onset of a long, debilitating, inevitably fatal disease. Any offspring of a parent with the gene is at 50 percent risk of inheriting it, and because symptoms of Huntington's do not generally appear until after child-bearing age, the disease is passed on to succeeding generations.

Before the development of the test, young adults from Huntington's families, like Nancy Wexler, often chose not to marry, or if they did, not to have children. But the arrival of the Huntington's test gave them an option, while posing a dilemma. If they decided to have the test and got negative results, they could breathe easy, secure in the knowledge that neither they nor any offspring would ever be stricken by the disorder. But a positive result, though it would end the uncertainty, could be devastating.

"Before the test was developed," says Wexler, "I always thought I'd want to know. It's knowledge macho. You're supposed to want to know everything." But given the reality of the test, she concedes, "you have to stop and say, 'On a day-to-day level, what impact is it going to have on me? What changes am I going to have to make in my life?' Denial in certain circumstances can be very healthy."

People whose test results confirm that they carry the Huntington's gene have become severely depressed, unable to lead normal lives. Even for those who seem to cope well with the grim knowledge, says Dorene Markel, director of the Family Collection Core at the University of Michigan's Human Genome Center, anticipation of the first signs of the disorder could color every aspect of life. "How are you going to look at stumbling on the street," she says, "or dropping a pencil? It's looking into a crystal ball and seeing the future. Medicine has never done this before."

Huntington's testing presents other, more subtle pitfalls. If a pregnant woman from a Huntington's family has never been tested but wants a prenatal test for her fetus, she could be dealt a double blow. If the fetus tests positive, it will confirm that not only the child but she, too, carries the Huntington's gene.

For all these reasons, geneticists and psychologists agree, extensive counseling must accompany any tests for serious genetic diseases. Indeed, preliminary reports indicate that professional counseling

IN LINE

▼

Right now, new nonionic contrast media and new imaging agents for MRI and nuclear medicine are moving through the R&D pipeline at Squibb Diagnostics.

THEN AND NOW

▼

In 1956, Squibb Diagnostics introduced the first iodinated contrast
medium: Renografin® (diatrizoate
acid meglumine/sodium). In 1986,
Squibb launched Isovue® (iopamidol injection), the nonionic with the
best iodine-to-carrier ratio. In 1991,
Squibb Diagnostics' tradition of research continues with new agents
being developed for CT, MRI, and
nuclear medicine.

makes a difference. A University of British Columbia study of individuals who had the Huntington's test and were found to be at increased risk concluded that "the earliest participants in a predictive testing program that is supported by extensive professional counseling have shown little evidence of severe psychological stress."

The fact that only the Huntington's markers, not the gene itself, have been identified further complicates testing and ethical considerations. Once a gene is isolated and sequenced, only the DNA of the person to be tested is needed, because the disease genes are virtually identical in all people. But if the test is for nearby markers, as is the case for Huntington's, Duchenne muscular dystrophy, and retinoblastoma, among others, blood samples must initially be taken from all members of the immediate family and from as many close relatives as possible. The reason: Each family has its characteristic marker, and it can be associated with those who have the disease only by taking blood samples and analyzing the family pedigree.

Consequently, if the person requesting the test is the first in a Huntington's family to do so, Markel explains, everyone has to consent to give blood. That can present ethical difficulties for physicians. Although each relative who donates blood is entitled to any genetic information the doctor learns about him or her, no one has the right to know about the DNA of the others.

The problem, explains geneticist Aubrey Milunsky, is that in the course of the test "you have to explain why you need the parents' blood and the aunts' and the uncles' and how it all works. Then after the test results are known, the person being tested can, by inference, figure out if Aunt Jane has the disease or not." For that reason and others—family feuds, for example—relatives often refuse to give blood, making testing difficult, if not impossible.

In administering tests, physicians can encounter other situations that sorely try their judgment and ethical standards. Dr. Milunsky cites the example of a pregnant woman who wanted prenatal testing of her fetus for a disease gene prevalent in her husband's family. After helping to collect blood samples from her husband and his brothers, sisters, aunt, and uncles, Dr. Milunsky says, "we did the tests and we discovered there was nonpaternity—in other words, the husband was not the father. The question is, what do you do with that? Call her into the office and say, 'I've got some good news and some bad news'? That is very difficult stuff, and a common problem."

Prenatal diagnosis for genetic disorders is particularly disturbing to pro-life activists, who liken it to search-and-destroy missions. In 1987 the Catholic Church included the procedure in a warning against high-tech medical practices affecting birth and reproduction. "Prenatal diagnosis must not be the equivalent of a death sentence," the Vatican declared. "A woman would be committing a gravely illicit act if she were to request a diagnosis with the deliberate intention of having an abortion should the results confirm the existence of a malformation or abnormality."

Even among those who favor a woman's right to have an abortion, the increasing rate of genetic discoveries and techniques raises questions. Though many prospective parents would not hesitate to seek an abortion if prenatal diagnosis confirmed such severe disorders as Tay-Sachs disease or spina bifida, positive test results for less serious or late-onset diseases could present them with a more difficult choice.

California representative Henry Waxman, a member of the congressional Biomedical Ethics Board, gives an example: "What do you do when you can measure a gene in the fetus which indicates that the person may get Alzheimer's disease at the age of 50? Do you terminate the fetus? What policies should we have?"

The same kinds of questions apply to neurofibromatosis (NF). This disorder causes internal and external nonmalignant tumors that frequently appear as barely noticeable coffee-colored blotches on the skin; but sometimes they are terribly disfiguring and, when they invade the central nervous system, can cause paralysis and death. Dr. Francis Collins points out that two-thirds of all NF victims live relatively normal lives "with punctuated moments of crisis." The other third, he says, "develop major problems that drastically alter their lives."

In July 1990, Dr. Collins and Ray White simultaneously announced that their groups had isolated the NF gene, which raised hopes that a test for the gene could be developed. But it will not be an easy task. The complex gene consists of a series of small exons, or coding regions, spread over at least a 200,000-base-pair stretch of chromosome 17, and both groups anticipate extensive research to further characterize the gene before a reliable test can be devised.

Even after a test is available, however, many questions will remain unresolved. It will not predict which of the highly variable consequences of the disease will occur, or when. "Even within the same family there is enormous variability in the severity of the disease," says Dr. Collins. "Prenatal diagnosis . . . will only predict whether the fetus is affected, not whether the disease will be mild or severe."

Nancy Wexler worries about the stigma that might be attached to the bearer of a disease gene, even when it is recessive. The millions of people who unknowingly carry a single copy of the CF gene suffer no ill effects and should be concerned only if they marry someone who also has the trait. But Wexler fears that when large-scale testing for the CF gene begins, those identified as carriers might be affected psychologically and perhaps also be viewed differently by society. "Would there be subtle or overt forms of labeling?" she wonders.

That was the case with sickle cell anemia when a nongenetic hemoglobin test for the disease was developed in the early 1970s. Because the sickle cell gene is also recessive, people carrying only one copy are generally not affected. Nonetheless, development of the sickle cell test was immediately followed by widespread discrimination against the estimated three million carriers, virtually all of whom are black. Blood samples from blacks seeking employment or volunteering for military service were routinely tested for sickle cell anemia, usually without the consent of the donor. Not only those with the disease but also those with the trait were often denied employment, charged higher insurance rates, barred from entry into the U.S. Air Force Academy, and otherwise discriminated against.

Although most of the restrictions against those with the sickle cell trait have since been abandoned or legally banned, fears have arisen that the proliferation of disease-gene tests might bring on a new wave of discriminatory practices. That the possibility exists has been dramatically shown by across-the-board discrimination against victims of AIDS. Indeed, Wexler's ethics group is using the reaction to the AIDS epidemic as a model. "One of the subjects we're discussing," she says, "is whether the legislation now in force to protect confidentiality for AIDS patients is relevant to genetic information, and how genetic information might affect employment, insurance, the criminal justice system, schools, and adoption agencies."

Civil libertarians worry that as more and more of the genome is interpreted, government agencies, employers, or insurance companies may begin to require and store genetic information on individuals without their consent. They envision giant data banks, equivalent to the FBI's fingerprint collection, that could bare the genetic flaws and predisposition of any citizen to prying eyes.

Their fears have been heightened by the use of DNA "fingerprinting," a technique based on the fact that no two genomes, except those of identical twins, are exactly alike. Employing gel electrophoresis, forensics experts can use the DNA from a bloodstain, a scrap of skin, semen, or even a single hair to distinguish the RFLP pattern of an individual from that of anyone else. Most courts now accept DNA fingerprinting as evidence in criminal cases, and a few states already legally store genetic data extracted from the DNA of convicted rapists and child molesters.

DNA fingerprinting has been used to obtain convictions in homicide and rape cases and to establish paternity. But it has not been the surefire forensic tool originally foreseen by police officials and prosecutors. In some recent cases, DNA-fingerprint evidence has been challenged, found wanting, and disallowed by the courts. The problem, says molecular geneticist Jan Witkowski of Cold Spring Harbor Laboratory, is that "no scientific technique is 100 percent reliable." Witkowski points out that Southern blotting is a long, tedious procedure. "A little bit can go wrong at each stage," he says, "and if not caught, this adds up at the end. DNA fingerprinting as a procedure is difficult to control."

Still, once the procedure is properly regulated, Witkowski insists, "it has a lot to offer forensic science." And, he adds, "it is not too much different from being identified by a Social Security number or by your picture on your driver's license." Baylor's Dr. Thomas Caskey agrees. He sees no ethical conflict or violation of privacy in the data-bank storage of DNA fingerprints. "All of the genetic markers that are currently being used in forensics," he says, "are taken from highly variable, informative regions generally located in non-gene areas of the genome." The markers, he stresses, are not associated with ethnic traits, medical traits, or physical characteristics and are used solely for identification. In August 1990, after doubts had arisen over the accuracy of DNA fingerprinting, a congressional study found that the technique was valid and reliable when properly performed.

If diagnosis for disease genes becomes commonplace, however, and if individual genetic profiles become as available as, say, credit ratings, ordinary citizens could suffer. "We would be creating a class of people who would find it much harder to find jobs or get insurance because of their risks," says Johns Hopkins physician Neil Holtzman. "It really raises concerns about what happens to those people who are at risk."

Pressures by employers and insurance companies to acquire genetic information will almost certainly mount as more of the genome is deciphered. The use of DNA testing in hiring to screen out those with potentially disabling genetic flaws could substantially lower a company's costs of unemployment compensation, medical insurance, pensions, and job training. It could also engender ill-informed, unwarranted discrimination against many capable employees or job applicants. Only a few U.S. companies now encourage or enforce genetic testing, though the number will surely grow.

Insurance companies, meanwhile, are already showing signs of nervousness. "The insurance industry is very concerned about this," says Dr. Holtzman, who in his book *Proceed With Caution* (Johns Hopkins University Press, 1989) examines the pitfalls of genetic testing. "If a policyholder knows about his genetic predisposition to a serious disease but the insurance company is forbidden from gathering that information, it makes private health insurance less and less viable."

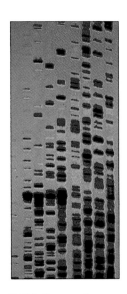

DNA fingerprints like the one above are accepted as evidence of identity by many courts.

The AIDS crisis has already driven that point home. Several states and the District of Columbia have forbidden insurance companies to test applicants for the AIDS virus or at least limited their ability to do so. However well-meaning that legislation was, it has resulted in what the industry calls "adverse selection," which occurs when a policy-holder, knowing that he or she is at increased risk for a serious or even fatal disease, loads up on health and life insurance. If the insurance company is unaware of that risk, it charges premiums that are far too low to cover the costs of eventual claims.

The problem may be magnified, Dr. Holtzman believes, as genetic diagnosis becomes available for susceptibility to widespread disorders such as heart disease and cancer. "Current thinking is that your genome is your private property," he says, "and it may be that we'll pass a law and say that insurance companies are not privy to this information—they can't ask the individual, they can't test for it directly, they can't ask the attending physician." That may be the ethical course, Dr. Holtzman says, but should such a law come to pass, it might spell the end of the private insurance industry. Then, he says, "you'll need some form of social insurance to protect the population, like national health insurance."

Economics and ethics also come into conflict when parents discover through prenatal testing that a fetus has serious genetic flaws that will burden them with crushing medical expenses after their child is born. Should they abort the fetus? Should an insurance company be required to pay medical bills? "If a couple feel that it is more consistent with their values to raise a child with a severe disease than to terminate the pregnancy," Dr. Holtzman says, "then that has to be honored at all costs. When you start to impose societal values and say that the cost-benefit ratio is what counts, then you start down a very treacherous, slippery slope."

He cites as an example a pregnant Louisiana woman from a family with cystic fibrosis who was asked by her health-care organization to undergo a prenatal test for the disease. If the fetus was affected, she was warned, the organization would not pay for the cost of caring for the child. "That constitutes tremendous pressure on the couple to terminate the pregnancy," says Dr. Holtzman. "It's an example of what I call a new eugenics."

Improper use of the new genetic information is also a concern of Albert Jonsen, a University of Washington ethicist. Until now, he says, doctors and researchers in the United States have confined themselves to uses of genetics for medical purposes. But, he asks, "will we stop with the medical applications, with the potential research or elimination of obvious, manifest diseases, or will we move into eugenics?"

Eugenics is a word often heard in discussions about the merits of the Human Genome Project. It's sometimes used by opponents of genetic engineering as a way to conjure images of Nazi Germany's race-purification programs. Knowledge of the genome, they fear, will lead to ill-advised attempts to improve it. But eugenics, loosely defined as the science of improving the inborn or hereditary qualities of a race or breed, can also have good connotations.

In a 1986 document entitled "Genetic Science for Human Benefit," the National Council of Churches supported genetic screening as a way to delete harmful genes from the gene pool. While conceding that some eugenic sciences have been "coercive, political, or diabolical, calling for mandatory sterilization, or even killing of 'unworthy' people," the document concludes that such measures as "appealing to men and women to abstain from procreation when they know they are carriers of disease" are "respectable" and "idealistic."

Theological support has also been expressed for the practice of gene therapy on humans. At a 1982 meeting of scientists at the Vatican, Pope John Paul II proclaimed that "the research of modern biology gives hope that the transfer and mutation of genes can ameliorate the condition of those who are affected by chromosomic diseases." And the National Council of Churches has not even ruled out what has become a bugaboo to many people, including some scientists: genetic therapy involving sex, or germ-line, cells. "If ever practicable," the council states, sex-cell therapy "will deserve especially stringent control. . . . Researchers and religious and secular commentators should approach it with extreme caution."

But the last word belongs to Thomas Murray, director of the Center for Biomedical Ethics at Case Western Reserve University. In defending the Human Genome Project, Murray addresses a thought that must fleetingly cross the minds of even some of the scientists involved in genetic research: that sequencing and deciphering the genome will reduce humans to little more than a string of four different chemical bases in various combinations and permutations, and that this ultimate exercise in reductionism will undermine their moral and spiritual standing and their dignity.

Nonsense, says Murray. "You know that somewhere there are squiggles in black ink that represent the notes to Beethoven's Ninth Symphony," he says, "but in no way does that diminish the grandeur of the symphony itself. And I don't think that the genome initiative diminishes the dignity of humankind. In fact, it may increase our appreciation for the Creator of all life. After all, Beethoven had 12 notes to work with, but the Creator had only four."

CHAPTER NINE

THE MEDICINE OF TOMORROW

The information flowing from the Human Genome Project will usher in the golden age of molecular medicine.

—Mark Pearson, director of molecular biology
at Du Pont Chemical Corporation

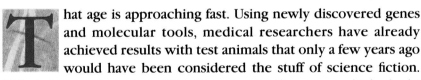

That age is approaching fast. Using newly discovered genes and molecular tools, medical researchers have already achieved results with test animals that only a few years ago would have been considered the stuff of science fiction. Although some technical difficulties remain, there seems little doubt that many of these remarkable procedures will move from the laboratory to the clinic within the next several years, bringing new hope to victims of serious hereditary disease.

"The genome project," says Dr. Leroy Hood, "is going to give us all the components of the very complicated physiological systems so that we can come to understand how they work together to make a heart or a kidney or a liver function." This detailed physiological knowledge, he says, will be the basis of a kind of molecular pharmacology. "We will rationally design drugs to be targeted to the particular protein that we know is effective in a particular case and help it carry out its function, or prevent it from killing a cell."

Dr. Hood's group at Caltech has already made substantial progress in animal tests. One experiment was prompted by the discovery of a protein that in laboratory mice stimulated certain immune-system cells to attack a mouse's own tissue, causing symptoms virtually identical to those of human multiple sclerosis (MS). By analyzing the protein, researchers determined what part of it was responsible for causing the autoimmune response. Then they synthesized a protein fragment with a structure similar to that of the MS protein, but different enough to prevent it from activating immune-system cells. When injected into the mouse, the synthetic protein "docked" with the immune cells and blocked their receptors, which are in effect molecular hookups for proteins. This blockade prevented the natural protein from engaging the immune cells, stimulating them, and causing the disease.

Dr. Hood's laboratory has also successfully experimented with a procedure that he says can eventually eliminate a genetic disorder from a family. Extracting a fertilized mouse egg that has grown to the 64-cell stage, the Caltech scientists remove several cells from the outer layer, extract the DNA, and mass-produce sequences of interest by polymerase chain reaction. Then they examine the DNA, searching for specific disease genes. "If it's defective," says Dr. Hood, "you throw the egg away. If it's good, you reimplant it in the mother, it grows to full term, and you get a healthy pup."

The same technique, Dr. Hood suggests, might someday be used in families with Huntington's disease. If a woman diagnosed as a carrier of the dominant Huntington's gene wants to have a baby she would opt for in vitro fertilization, using the procedure envisioned by Dr. Hood. Several of her eggs would be extracted and fertilized in a test tube with her husband's sperm. "Then you would work with the fertilized eggs until you found one that had no Huntington's genes and you would implant it," Dr. Hood says. "That way there's no abortion. You make the decision at the stage of the fertilized egg." In this family, the chain of disease would be broken because the resulting children would be free of the deadly gene.

That approach has already been used by doctors at Hammersmith Hospital in London to determine the sex of embryos of couples at risk of transmitting to their male offspring diseases caused by genes on the X chromosome. Single cells from six- to eight-cell embryos conceived in a test tube are snipped off and their DNA analyzed for a unique segment of the Y chromosome. Embryos found to be male are discarded, but those with two X chromosomes are implanted. The procedure has resulted in eight successful implantations and four births.

As remarkable as these techniques are, they pale by comparison with the highly touted and still controversial technique of human gene therapy—the introduction of normal genes into existing cells to cure hereditary diseases. In light of successful experiments with plants and animals, gene therapy seems reasonable enough.

READING GLASSES

▼

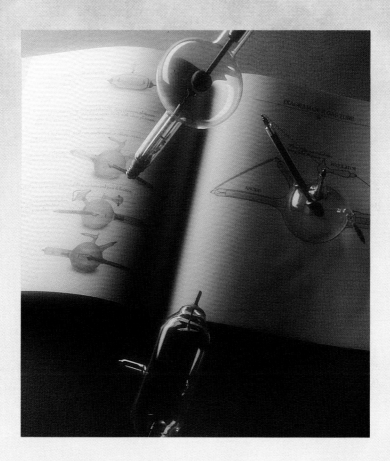

Squibb Diagnostics has recently published a limited edition of Dr. William Shehadi's *Reflections on the Radiology Years*. This new book serves as a guide to Dr. Shehadi's vast collection of glass-bulb x-ray tubes, now on display at the Johns Hopkins University School of Medicine.

ISOVUE® *...enhances your image*
(iopamidol injection)

RAISING RADIOLOGY

Squibb Diagnostics is a Vanguard Contributor to the Radiological Society of North America, providing educational support for the next generation of radiologists.

SQUIBB™
Diagnostics

A Bristol-Myers Squibb Company

Still, researchers have encountered formidable problems, and rosy predictions that the technique would be successfully used on humans by the mid-1980s proved to be overly optimistic. In September 1990, however, shortly after the NIH and the Food and Drug Administration had given their final approval, molecular biologist W. French Anderson and immunologist Michael Blaese, both with the NIH, conducted the first authorized test of human gene therapy. Their patient was a 4-year-old girl suffering from adenosine deaminase (ADA) deficiency, a rare disease afflicting children born without the gene that codes for ADA, an enzyme that breaks down harmful metabolic byproducts. Without the enzyme, these byproducts accumulate in the body and destroy T cells and B cells, inactivating the immune system.

Nine days before the procedure, T cells were removed from the child's body and exposed to mouse leukemia retroviruses equipped with human ADA genes spliced into their genomes and rendered harmless by genetic engineering. The retroviruses invaded the human T cells and deposited their genetic material (including the ADA gene) into the DNA of the cells, which promptly began churning out the ADA enzyme. In their historic half-hour procedure, the NIH researchers dripped a solution containing a billion or so of the gene-altered T cells into a vein in the child's left hand. Physicians plan to monitor her carefully over the next few years to determine if the altered T cells are supplying her with the vital enzyme and to administer monthly infusions of new cells to replace those that have died. As an added precaution, the girl is to continue treatment with PEG-ADA, a drug with a chemically altered form of the missing enzyme that she had been taking almost since infancy.

In February 1991, Dr. Blaese reported that the child was responding to therapy. "We're starting to see improved immune function," he said. Even if it works, however, this form of gene therapy will at best be only a holding action against ADA deficiency disease. Dr. Anderson and Dr. Blaese hope to effect an actual cure by inserting the ADA gene into bone-marrow cells, which would then provide a continuous supply of T cells capable of producing the enzyme.

It is no coincidence, says Richard Mulligan, a leading gene therapy researcher at MIT, that many of the disorders initially targeted for human gene therapy involve defective genes expressed in bone-marrow cells. "Everyone had thought about the bone-marrow system for gene therapy," he says, "because there is a well-characterized method for doing the transplant. It's often done on cancer patients."

Building on this experience, gene therapists have evolved a strategy for conquering such blood diseases as beta-thalassemia, caused by a defective gene in bone-marrow cells. Deprived of the protein coded

Dr. W. French Anderson, along with Dr. Michael Blaese, initiated the first authorized gene therapy test on a child with adenosine deaminase (ADA) deficiency. Says Dr. Anderson: "We're cautiously optimistic."

for by the gene, hemoglobin cannot function properly, which results in a severe form of anemia.

Using techniques successful in animal tests, researchers will extract bone-marrow cells from a beta-thalassemia victim, place them in a laboratory dish, and expose them to retroviruses engineered to carry correctly functioning versions of the patient's faulty gene. When a retrovirus invades a marrow cell, it will insert itself into the cell's DNA, carrying the good gene with it. Reimplanted in the marrow, the altered marrow cells will take hold and multiply, churning out the previously lacking protein and relieving the symptoms of beta-thalassemia.

In any of these procedures, there would seem to be risk in the use of a retrovirus, which consists largely of a single strand of RNA and a little protein. Some retroviruses, after all, are known to cause serious diseases such as adult T cell leukemia-lymphoma and AIDS. But gene therapy retroviruses have been sanitized. Using restriction enzymes, scientists have snipped out the segments of RNA that enable the retroviruses to multiply and do their mischief, reducing the virus to a convenient vector for transporting the good gene into cellular DNA. That transformation, says molecular geneticist Theodore Friedmann, of the University of California at San Diego, converts "the swords of pathology into the plowshares of therapy."

Inserting the proper cargo into the retrovirus, however, is still far from a cut-and-dried procedure. For the gene to be properly expressed once it has been inserted into the genome of a patient, it must be accompanied by the DNA segments that mark its beginning and end and tell it when and how often to turn on and off. In one experiment with laboratory rats, for example, a transplanted gene that coded for growth hormone functioned excessively, and an experimental rat grew to a giant size. But as scientists isolate more genes and decipher the adjoining control and signal sequences, many of those difficulties should be overcome.

Medical researchers considering gene therapy for cells other than those in the marrow face greater obstacles. MIT's Mulligan has been looking into the possibility of curing hypercholesterolemia, a condition caused by defective receptors in liver cells. These inoperative receptors, resulting from a faulty gene, prevent the liver from removing cholesterol from the blood, processing it, and passing it along to be eliminated from the body in urine. Mulligan's goal is to incorporate a good gene into liver cells extracted from a patient and then to re-implant the genetically modified cells in the liver, where—at least in theory—they will persist and function. "Introduction of the genetically modified cells is relatively straightforward," Mulligan says. "The difficulty is to obtain their persistence after transplantation."

Another concern is the possibility of unforeseen consequences that might arise if the retrovirus inserts itself into the wrong place in the genome and disrupts the normal functioning of another gene, causing cancer or other disorders. An additional drawback of retroviruses is their limited cargo capacity. "Some genes," says Mulligan, "are too big to fit into a retrovirus vector," but scientists are already working on alternative carriers for large genetic loads.

Despite such worries, experimenters have scored many gene therapy successes with laboratory animals and have actually cured some hereditary diseases. At Caltech, for example, researchers tried gene therapy on so-called shiverer mice, which are born with a defective

Dr. Steven Rosenberg is attempting to treat two terminal melanoma patients with genetically engineered white blood cells carrying the human gene that codes for tumor necrosis factor.

nervous-system gene that causes them to shake continuously, have frequent convulsions, and die early. The Caltech scientists cloned normal versions of the nervous-system gene and, using a kind of shotgun approach, injected them into fertilized shiverer-mice eggs, which were then reimplanted in the mothers. In about one in 150 of these embryos, the new gene functioned properly and a normal mouse was born.

More precise methods of mechanically injecting genes into DNA are under investigation. "In another 10 years or so," Dr. Hood predicts, "we'll be able to target genes to just the right place." The trick, he says, is to attach to the gene flanking sequences of nucleotides that are complementary to those on either side of the point in the genome where the gene should be inserted. "The tendency is for the flanking sequences to look throughout the genome and match up in a complementary way, carrying the gene with them," he explains. "So the sequences take the gene to where it should be inserted."

The first known attempt at human gene therapy occurred in 1980. It was performed by two UCLA scientists who, despairing of their chances of winning approval for their experiment from the complicated NIH bureaucracy, traveled to Italy and Israel to treat patients suffering from beta-thalassemia. Using engineered retroviruses carrying normal human genes, they infected marrow cells extracted from the victims and reimplanted them. But their techniques were still not perfected, the transplants failed, and the disease eventually claimed the lives of the victims. The NIH censured the lead scientist and then penalized him by withholding certain grant monies from him for several years.

That unhappy incident marked the last known attempt at transferring new genes into humans until May 22, 1989, when Dr. Anderson and surgeon Steven Rosenberg, also with the National Institutes of Health, performed a groundbreaking experiment. Using a technique approved beforehand by the NIH's Recombinant DNA Advisory Committee, they injected genetically engineered white blood cells into the first of 10 patients suffering from advanced melanoma. The cells, called tumor-infiltrating lymphocytes (TILs), had been altered by the insertion of a bacterial gene so that they could serve as markers for researchers trying to track them through the body. Dr. Anderson and Dr. Rosenberg were attempting to shed light on the results of an earlier test of melanoma patients. In that study, Dr. Rosenberg had extracted TILs from the tumors of 20 melanoma victims and exposed them to interleukin 2, a natural substance that energizes the lymphocytes, making them more voracious cancer-cell killers. Then he injected the activated TILs back into the bloodstreams of the melanoma patients and a few months later reported dramatic results: tumors had regressed in 11 of the 20 test participants.

Why had the unique therapy worked for some patients and not others? The 1989 test was designed to provide an answer. This time, the TILs extracted from the patients and activated by interleukin 2 were exposed to a mouse leukemia retrovirus that had been significantly reengineered. Sequences in its genome that enable it to reproduce had been snipped out and a bacterial gene inserted. The gene was selected because it codes for a protein that neutralizes an antibiotic capable of killing human cells.

As expected, the retrovirus invaded the TILs, endowing them with the bacterial gene; the TILs were injected back into the melanoma patients. This time, however, Dr. Rosenberg could follow the progress of the cells as they moved in to attack the tumors. By exposing blood and tumor samples to the cell-killing antibiotic, he destroyed all but the altered TILs, which were then easily detectable. In September 1989, Dr. Rosenberg was able to announce that the TILs had begun to concentrate in the tumors five days after they were injected, and that they were still viable in the bloodstream after 19 days.

The success of that experiment led, in January 1991, to a second test of gene therapy on humans, a 29-year-old woman and a 42-year-old man, both with terminal melanoma. Dr. Rosenberg infused each of the patients with about 100 million TILs that had been extracted from their tumors and reengineered. But this time the TILs had been modified, via retroviruses, to carry a human gene that codes for tumor necrosis factor (TNF), a naturally occurring compound that attacks cancer cells. Dr. Rosenberg's expectations were that the TILs would home in on the tumors, attacking the cancerous cells and releasing the anti-tumor factor to help finish them off. Plans called for the patients to receive additional infusions, in gradually larger doses, every three weeks, while doctors monitored their progress.

This and other proposed forms of gene therapy deal solely with correcting genetic deficiencies in the somatic, or non-sex, cells. But if an enzymatic deficiency is cured, say, by introducing a normal gene into liver cells, the patient's genome remains unchanged; all of the other body cells, including the gene cells, retain the defective gene, which can then be passed on to subsequent generations. The question then arises: Why not simply perform gene therapy on the sex cells—the sperm or the egg? That would change the message of the genome by excising the genetic defect even before the embryo is conceived by the union of the egg and sperm. Even more significant, in the words of Daniel Koshland, a biochemist at the University of California at Berkeley, "it can reverse a family's history by ending the progress of a deficient gene into new generations."

Yet it is just that ability that disturbs some scientists and many nonscientists. The introduction of new genes (and the excision of others) in germ cells would be duplicated in every cell of the progeny, including their germ cells. If a new gene were inserted at the wrong location, for example, or with incomplete or faulty accompanying control sequences, the results could be disastrous for the offspring. The functioning of other genes nearby could be disrupted. They might be activated prematurely, causing cancer or other disorders, or be turned off permanently, cutting off the production of a vital enzyme. Other unpredictable consequences might occur.

"I think that before we do germ-cell manipulations," says Dr. Hood, "we should understand very, very much more about the implications of what we're doing in the long term, in evolutionary terms." He cites the fact that genes contributed by the mother sometimes seem to have a greater effect in offspring than the equivalent genes from the father—a phenomenon that escaped the attention of Mendel and, until recently, contemporary geneticists. "I don't find that surprising at all," Dr. Hood says. "The maternal egg is thousands of times bigger than the sperm, and it has all the mitochondria. That's something that evolution has taken two billion years to work out, and there's presumably a logical reason for things being that way."

Dr. Hood would caution researchers who might want to try to balance out the maternal and paternal influences. "I would say, 'Hey, wait a minute. We don't begin to understand the reason for this difference, and if we manipulate it for a superficial reason, we may end up creating a horrible mess.' That is precisely the reason we shouldn't fool with the germ line, because we don't yet know about the consequences."

In an attempt to learn more about those consequences, scientists are proceeding with experiments that involve manipulating the genes in the germ cells of laboratory animals. Most of them expect that human germ cells, too, will eventually be modified to conquer disease. Though that day may be far off, the information and insights flowing from the genome project are bound to speed its arrival.

Even when the entire sequence has been read and all of the human genes identified and located, the job of deciphering many of heredity's subtle messages will remain. For hidden in the genome's three billion code letters are clues to evolution, to human development, and perhaps to many of the still-mysterious workings of the brain. Norton Zinder summed it up in January 1989 as he opened the first meeting of the Human Genome Advisory Committee: "Today we begin. We are instituting an unending study of human biology. Whatever it's going to be, it will be an adventure, a priceless endeavor. And when it's done, someone else will sit down and say, 'It's time to begin.'"

Additional Copies

To order additional copies of *The New Genetics*
for friends or colleagues, please write to The Grand
Rounds Press, Whittle Direct Books, 505 Market St.,
Knoxville, Tenn. 37902. Please include the recipient's
name, mailing address, and, where applicable, primary
specialty and ME number.

For a single copy, please enclose a check for $21.95,
plus $1.50 for postage and handling, payable to The
Grand Rounds Press. When ordering 10 or more books,
enclose $20.95 for each, plus $5 for postage and
handling. For orders of 50 or more books, enclose
$19.95 for each, plus $20 for postage and handling. For
more information about The Grand Rounds Press,
please call 800-765-5889.

Also available, at the same prices, are copies of the
previous book from The Grand Rounds Press, *The
Doctor Watchers* by Spencer Vibbert.

Please allow four weeks for delivery.
Tennessee residents must add 7¾ percent sales tax.

ADVERTISER'S APPENDIX

ISOVUE®-200 Iopamidol Injection 41%
ISOVUE®-300 Iopamidol Injection 61%
ISOVUE®-370 Iopamidol Injection 76%

DIAGNOSTIC NONIONIC RADIOPAQUE CONTRAST MEDIA
For Angiography Throughout the Cardiovascular System, Including Cerebral and Peripheral Arteriography, Coronary Arteriography and Ventriculography, Pediatric Angiocardiography, Selective Visceral Arteriography and Aortography, Peripheral Venography (Phlebography), and Adult and Pediatric Intravenous Excretory Urography and Intravenous Adult and Pediatric Contrast Enhancement of Computed Tomographic (CECT) Head and Body Imaging

DESCRIPTION

ISOVUE (Iopamidol Injection) formulations are stable, aqueous, sterile, and nonpyrogenic solutions for intravascular administration.

Each mL of ISOVUE-200 (Iopamidol Injection 41%) provides 408 mg iopamidol with 1 mg tromethamine and 0.26 mg edetate calcium disodium. The solution contains approximately 0.029 mg (0.001 mEq) sodium and 200 mg organically bound iodine per mL.

Each mL of ISOVUE-300 (Iopamidol Injection 61%) provides 612 mg iopamidol with 1 mg tromethamine and 0.39 mg edetate calcium disodium. The solution contains approximately 0.043 mg (0.002 mEq) sodium and 300 mg organically bound iodine per mL.

Each mL of ISOVUE-370 (Iopamidol Injection 76%) provides 755 mg iopamidol with 1 mg tromethamine and 0.48 mg edetate calcium disodium. The solution contains approximately 0.053 mg (0.002 mEq) sodium and 370 mg organically bound iodine per mL.

The pH of ISOVUE contrast media has been adjusted to 6.5-7.5 with hydrochloric acid. Pertinent physicochemical data are noted below. ISOVUE (Iopamidol Injection) is hypertonic as compared to plasma and cerebrospinal fluid (approximately 285 and 301 mOsm/kg water, respectively).

Parameter	41%	Iopamidol 61%	76%
Concentration (mgI/mL)	200	300	370
Osmolality @ 37°C (mOsm/kg water)	413	616	796
Viscosity (cP) @ 37°C	2.0	4.7	9.4
@ 20°C	3.3	8.8	20.9
Specific Gravity @ 37°C	1.216	1.328	1.405

Iopamidol is designated chemically as (S)-N,N'-bis[2-hydroxy-l-(hydroxymethyl)-ethyl]-2,4,6-triiodo-5-lactamidoisophthalamide.

CLINICAL PHARMACOLOGY

Intravascular injection of a radiopaque diagnostic agent opacifies those vessels in the path of flow of the contrast medium, permitting radiographic visualization of the internal structures of the human body until significant hemodilution occurs.

Following intravascular injection, radiopaque diagnostic agents are immediately diluted in the circulating plasma. Calculations of apparent volume of distribution at steady-state indicate that iopamidol is distributed between the circulating blood volume and other extracellular fluid; there appears to be no significant deposition of iopamidol in tissues. Uniform distribution of iopamidol in extracellular fluid is reflected by its demonstrated utility in contrast enhancement of computed tomographic imaging of the head and body following intravenous administration.

The pharmacokinetics of intravenously administered iopamidol in normal subjects conform to an open two-compartment model with first order elimination (a rapid alpha phase for drug distribution and a slow beta phase for drug elimination). The elimination serum or plasma half-life is approximately two hours; the half-life is not dose dependent. No significant metabolism, deiodination, or biotransformation occurs.

Iopamidol is excreted mainly through the kidneys following intravascular administration. In patients with impaired renal function, the elimination half-life is prolonged dependent upon the degree of impairment. In the absence of renal dysfunction, the cumulative urinary excretion for iopamidol, expressed as a percentage of administered intravenous dose, is approximately 35 to 40 percent at 60 minutes, 80 to 90 percent at 8 hours, and 90 percent or more in the 72- to 96-hour period after administration. In normal subjects, approximately one percent or less of the administered dose appears in cumulative 72- to 96-hour fecal specimens.

ISOVUE may be visualized in the renal parenchyma within 30-60 seconds following rapid intravenous administration. Opacification of the calyces and pelves in patients with normal renal function becomes apparent within 1 to 3 minutes, with optimum contrast occurring between 5 and 15 minutes. In patients with renal impairment, contrast visualization may be delayed.

Iopamidol displays little tendency to bind to serum or plasma proteins.

No evidence of *in vivo* complement activation has been found in normal subjects.

Animal studies indicate that iopamidol does not cross the blood-brain barrier to any significant extent following intravascular administration.

ISOVUE (Iopamidol Injection) enhances computed tomographic brain imaging through augmentation of radiographic efficiency. The degree of enhancement of visualization of tissue density is directly related to the iodine content in an administered dose; peak iodine blood levels occur immediately following rapid injection of the dose. These levels fall rapidly within five to ten minutes. This can be accounted for by the dilution in the vascular and extracellular fluid compartments which causes an initial sharp fall in plasma concentration. Equilibration with the extracellular compartments is reached in about ten minutes; thereafter, the fall becomes exponential. Maximum contrast enhancement frequently occurs after peak blood iodine levels are reached. The delay in maximum contrast enhancement can range from five to forty minutes depending on the peak iodine levels achieved and the cell type of the lesion. This lag suggests that radiographic contrast enhancement is at least in part dependent on the accumulation of iodine within the lesion and outside the blood pool, although the mechanism by which this occurs is not clear. The radiographic enhancement of nontumoral lesions, such as arteriovenous malformations and aneurysms, is probably dependent on the iodine content of the circulating blood pool.

In CECT head imaging, ISOVUE (Iopamidol Injection) does not accumulate in normal brain tissue due to the presence of the "blood-brain" barrier. The increase in x-ray absorption in normal brain is due to the presence of contrast agent within the blood pool. A break in the blood-brain barrier such as occurs in malignant tumors of the brain allows the accumulation of the contrast medium within the interstitial tissue of the tumor. Adjacent normal brain tissue does not contain the contrast medium.

In nonneural tissues (during computed tomography of the body), iopamidol diffuses rapidly from the vascular into the extravascular space. Increase in x-ray absorption is related to blood flow, concentration of the contrast medium, and extraction of the contrast medium by interstitial tissue of tumors since no barrier exists. Contrast enhancement is thus due to the relative differences in extravascular diffusion between normal and abnormal tissue, quite different from that in the brain.

The pharmacokinetics of iopamidol in both normal and abnormal tissue have been shown to be variable. Contrast enhancement appears to be greatest soon after administration of the contrast medium, and following intraarterial rather than intravenous administration. Thus, greatest enhancement can be detected by a series of consecutive two- to three-second scans performed just after injection (within 30 to 90 seconds), i.e., dynamic computed tomographic imaging.

INDICATIONS AND USAGE

ISOVUE (Iopamidol Injection) is indicated for angiography throughout the cardiovascular system, including cerebral and peripheral arteriography, coronary arteriography and ventriculography, pediatric angiocardiography, selective visceral arteriography and aortography, peripheral venography (phlebography), and intravenous excretory urography and intravenous adult and pediatric contrast enhancement of computed tomographic (CECT) head and body imaging (see below).

CECT Head Imaging

ISOVUE may be used to refine diagnostic precision in areas of the brain which may not otherwise have been satisfactorily visualized.

Tumors

ISOVUE may be useful to investigate the presence and extent of certain malignancies such as: gliomas including malignant gliomas, glioblastomas, astrocytomas, oligodendrogliomas and gangliomas, ependymomas, medulloblastomas, meningiomas, neuromas, pinealomas, pituitary adenomas, craniopharyngiomas, germinomas, and metastatic lesions. The usefulness of contrast enhancement for the investigation of the retrobulbar space and in cases of low grade or infiltrative glioma has not been demonstrated.

In calcified lesions, there is less likelihood of enhancement. Following therapy, tumors may show decreased or no enhancement.

The opacification of the inferior vermis following contrast media administration has resulted in false-positive diagnosis in a number of otherwise normal studies.

Nonneoplastic Conditions

ISOVUE may be beneficial in the image enhancement of nonneoplastic lesions. Cerebral infarctions of recent onset may be better visualized with contrast enhancement, while some infarctions are obscured if contrast media are used. The use of iodinated contrast media results in contrast enhancement in about 60 percent of cerebral infarctions studied from one to four weeks from the onset of symptoms.

Sites of active infection may also be enhanced following contrast media administration.

Arteriovenous malformations and aneurysms will show contrast enhancement. For these vascular lesions, the enhancement is probably dependent on the iodine content of the circulating blood pool.

Hematomas and intraparenchymal bleeders seldom demonstrate any contrast enhancement. However, in cases of intraparenchymal clot, for which there is no obvious clinical explanation, contrast media administration may be helpful in ruling out the possibility of associated arteriovenous malformation.

CECT Body Imaging

ISOVUE (Iopamidol Injection) may be used for enhancement of computed tomographic images for detection and evaluation of lesions in the liver, pancreas, kidneys, aorta, mediastinum, abdominal cavity, pelvis and retroperitoneal space.

Enhancement of computed tomography with ISOVUE may be of benefit in establishing diagnoses of certain lesions in these sites with greater assurance than is possible with CT alone, and in supplying additional features of the lesions (e.g., hepatic abscess delineation prior to percutaneous drainage). In other cases, the contrast agent may allow visualization of lesions not seen with CT alone (e.g., tumor extension), or may help to define suspicious lesions seen with unenhanced CT (e.g., pancreatic cyst).

Contrast enhancement appears to be greatest within 60 to 90 seconds after bolus administration of contrast agent. Therefore, utilization of a continuous scanning technique ("dynamic CT scanning") may improve enhancement and diagnostic assessment of tumor and other lesions such as an abscess, occasionally revealing unsuspected or more extensive disease. For example, a cyst may be distinguished from a vascularized solid lesion when precontrast and enhanced scans are compared; the nonperfused mass shows unchanged x-ray absorption (CT number). A vascularized lesion is characterized by an increase in CT number in the few minutes after a bolus of intravascular contrast agent; it may be malignant, benign, or normal tissue, but would probably not be a cyst, hematoma, or other nonvascular lesion.

Because unenhanced scanning may provide adequate diagnostic information in the individual patient, the decision to employ contrast enhancement, which may be associated with risk and increased radiation exposure, should be based upon a careful evaluation of clinical, other radiological, and unenhanced CT findings.

CONTRAINDICATIONS

None.

WARNINGS

Nonionic iodinated contrast media inhibit blood coagulation, *in vitro,* less than ionic contrast media. Clotting has been reported when blood remains in contact with syringes containing nonionic contrast media.

Serious, rarely fatal, thromboembolic events causing myocardial infarction and stroke have been reported during angiographic procedures with both ionic and nonionic contrast media. Therefore, meticulous intravascular administration technique is necessary, particularly during angiographic procedures, to minimize thromboembolic events. Numerous factors, including length of procedure, catheter and syringe material, underlying disease state, and concomitant medications may contribute to the development of thromboembolic events. For these reasons, meticulous angiographic techniques are recommended including close attention to guidewire and catheter manipulation, use of manifold systems and/or three way stopcocks, frequent catheter flushing with heparinized saline solutions, and minimizing the length of the procedure. The use of plastic syringes in place of glass syringes has been reported to decrease but not eliminate the likelihood of *in vitro* clotting.

Caution must be exercised in patients with severely impaired renal function, those with combined renal and hepatic disease, or anuria, particularly when larger doses are administered.

Radiopaque diagnostic contrast agents are potentially hazardous in patients with multiple myeloma or other paraproteinemia, particularly in those with therapeutically resistant anuria. Myeloma occurs most commonly in persons over age 40. Although neither the contrast agent nor dehydration has been proved separately to be the cause of anuria in myelomatous patients, it has been speculated that the combination of both may be causative. The risk in myelomatous patients is not a contraindication; however, special precautions are required.

Contrast media may promote sickling in individuals who are homozygous for sickle cell disease when injected intravenously or intraarterially.

Administration of radiopaque materials to patients known or suspected of having pheochromocytoma should be performed with extreme caution. If, in the opinion of the physician, the possible benefits of such procedures outweigh the considered risks, the procedures may be performed; however, the amount of radiopaque medium injected should be kept to an absolute minimum. The blood pressure should be assessed throughout the procedure and measures for treatment of a hypertensive crisis should be available. These patients should be monitored very closely during contrast enhanced procedures.

Reports of thyroid storm following the use of iodinated radiopaque diagnostic agents in patients with hyperthyroidism or with an autonomously functioning thyroid nodule suggest that this additional risk be evaluated in such patients before use of any contrast medium.

PRECAUTIONS

General

Diagnostic procedures which involve the use of any radiopaque agent should be carried out under the direction of personnel with the prerequisite training and with a thorough knowledge of the particular procedure to be performed. Appropriate facilities should be available for coping with any complication of the procedure, as well as for emergency treatment of severe reaction to the contrast agent itself. After parenteral administration of a radiopaque agent, competent personnel and emergency facilities should be available for at least 30 to 60 minutes since severe delayed reactions may occur.

Preparatory dehydration is dangerous and may contribute to acute renal failure in patients with advanced vascular disease, diabetic patients, and in susceptible nondiabetic patients (often elderly with preexisting renal disease). *Patients should be well hydrated prior to and following iopamidol administration.*

The possibility of a reaction, including serious, life-threatening, fatal, anaphylactoid or cardiovascular reactions, should always be considered (see ADVERSE REACTIONS). Patients at increased risk include those with a history of a previous reaction to a contrast medium, patients with a known sensitivity to iodine per se, and patients with a known clinical hypersensitivity (bronchial asthma, hay fever, and food allergies). The occurrence of severe idiosyncratic reactions has prompted the use of several pretesting methods. However, pretesting cannot be relied upon to predict severe reactions and may itself be hazardous for the patient. It is suggested that a thorough medical history with emphasis on allergy and hypersensitivity, prior to the injection of any contrast medium, may be more accurate than pretesting in predicting potential adverse reactions. A positive history of allergies or hypersensitivity does not arbitrarily contraindicate the use of a contrast agent where a diagnostic procedure is thought essential, but caution should be exercised. Premedication with antihistamines or corticosteroids to avoid or minimize possible allergic reactions in such patients should be considered. Recent reports indicate that such pretreatment does not prevent serious life-threatening reactions, but may reduce both their incidence and severity.

General anesthesia may be indicated in the performance of some procedures in selected patients; however, a higher incidence of adverse reactions has been reported with radiopaque media in anesthetized patients, which may be attributable to the inability of the patient to identify untoward symptoms, or to the hypotensive effect of anesthesia which can reduce cardiac output and increase the duration of exposure to the contrast agent.

Even though the osmolality of iopamidol is low compared to diatrizoate or iothalamate based ionic agents of comparable iodine concentration, the potential transitory increase in the circulatory osmotic load in patients with congestive heart failure requires caution during injection. These patients should be observed for several hours following the procedure to detect delayed hemodynamic disturbances.

In angiographic procedures, the possibility of dislodging plaques or damaging or perforating the vessel wall should be borne in mind during catheter manipulations and contrast medium injection. Test injections to ensure proper catheter placement are suggested.

Selective coronary arteriography should be performed only in selected patients and those in whom the expected benefits outweigh the procedural risk. The inherent risks of *angiocardiography* in patients with chronic pulmonary emphysema must be weighed against the necessity for performing this procedure. *Angiography* should be avoided whenever possible in patients with homocystinuria, because of the risk of inducing thrombosis and embolism. See also Pediatric Use.

In addition to the general precautions previously described, special care is required when venography is performed in patients with suspected thrombosis, phlebitis, severe ischemic disease, local infection or a totally obstructed venous system.

Extreme caution during injection of contrast media is necessary to avoid extravasation and fluoroscopy is recommended. This is especially important in patients with severe arterial or venous disease.

Information for Patients
Patients receiving injectable radiopaque diagnostic agents should be instructed to:
1. Inform your physician if you are pregnant.
2. Inform your physician if you are diabetic or if you have multiple myeloma, pheochromocytoma, homozygous sickle cell disease, or known thyroid disorder (see WARNINGS).
3. Inform your physician if you are allergic to any drugs, food, or if you had any reactions to previous injections of substances used for x-ray procedures (see PRECAUTIONS, General).
4. Inform your physician about any other medications you are currently taking, including nonprescription drugs, before you have this procedure.

Drug Interactions
Renal toxicity has been reported in a few patients with liver dysfunction who were given oral cholecystographic agents followed by intravascular contrast agents. Administration of intravascular agents should therefore be postponed in any patient with a known or suspected hepatic or biliary disorder who has recently received a cholecystographic contrast agent.

Other drugs should not be admixed with iopamidol.

Drug/Laboratory Test Interactions
The results of PBI and radioactive iodine uptake studies, which depend on iodine estimations, will not accurately reflect thyroid function for up to 16 days following administration of iodinated contrast media. However, thyroid function tests not depending on iodine estimations, e.g., T3 resin uptake and total or free thyroxine (T4) assays are not affected.

Any test which might be affected by contrast media should be performed prior to administration of the contrast medium.

Laboratory Test Findings
In vitro studies with animal blood showed that many radiopaque contrast agents, including iopamidol, produced a slight depression of plasma coagulation factors including prothrombin time, partial thromboplastin time, and fibrinogen, as well as a slight tendency to cause platelet and/or red blood cell aggregation (see PRECAUTIONS-General).

Transitory changes may occur in red cell and leucocyte counts, serum calcium, serum creatinine, serum glutamic oxalacetic transaminase (SGOT), and uric acid in urine; transient albuminuria may occur.

These findings have not been associated with clinical manifestations.

Carcinogenesis, Mutagenesis, Impairment of Fertility
In animal reproduction studies performed on rats, intravenously administered iopamidol did not induce adverse effects on fertility or general reproductive performance.

In studies to determine mutagenic activity, iopamidol did not cause any increase in mutation rates.

Pregnancy Category B
No teratogenic effects attributable to iopamidol have been observed in teratology studies performed in animals. There are, however, no adequate and well controlled studies in pregnant women. It is not known whether iopamidol crosses the placental barrier or reaches fetal tissues. However, many injectable contrast agents cross the placental barrier in humans and appear to enter fetal tissues passively. Because animal teratology studies are not always predictive of human response, this drug should be used during pregnancy only if clearly needed.

Radiologic procedures involve a certain risk related to the exposure of the fetus to ionizing radiation.

Labor and Delivery
It is not known whether use of contrast agents during labor or delivery has immediate or delayed adverse effects on the fetus, prolongs the duration of labor or increases the likelihood that forceps delivery or other obstetrical intervention or resuscitation of the newborn will be necessary.

Nursing Mothers
It is not known whether iopamidol is excreted in human milk. However, many injectable contrast agents are excreted unchanged in human milk. Although it has not been established that serious adverse reactions occur in nursing infants, caution should be exercised when intravascular contrast media are administered to nursing women because of potential adverse reactions, and consideration should be given to temporarily discontinuing nursing.

Pediatric Use
Safety and effectiveness in children has been established in pediatric angiocardiography, computed tomography (head and body), and excretory urography. Pediatric patients at higher risk of experiencing adverse events during contrast medium administration may include those having asthma, a sensitivity to medication and/or allergens, cyanotic heart disease, congestive heart failure, a serum creatinine greater than 1.5 mg/dL or those less than 12 months of age.

ADVERSE REACTIONS
Adverse reactions following the use of iopamidol are usually mild to moderate, self-limited, and transient.

In angiocardiography (597 patients), the adverse reactions with an estimated incidence of one percent or higher are: hot flashes 3.4%; angina pectoris 3.0%; flushing 1.8%; bradycardia 1.3%; hypotension 1.0%; hives 1.0%.

In a clinical trial with 76 pediatric patients undergoing angiocardiography, 2 adverse reactions (2.6%) both remotely attributed to the contrast media were reported. Both patients were less than 2 years of age, both had cyanotic heart disease with underlying right ventricular abnormalities and abnormal pulmonary circulation. In one patient pre-existing cyanosis was transiently intensified following contrast media administration. In the second patient pre-existing decreased peripheral perfusion was intensified for 24 hours following the examination. (See "Precautions" Section for information on high risk nature of these patients.)

Intravascular injection of contrast media is frequently associated with the sensation of warmth and pain, especially in peripheral arteriography and venography; pain and warmth are less frequent and less severe with ISOVUE (Iopamidol Injection) than with diatrizoate meglumine and diatrizoate sodium injection.

The following table of incidence of reactions is based on clinical studies with ISOVUE in about 2191 patients.

Adverse Reactions
Estimated Overall Incidence

System	>1%	≤1%
Cardiovascular	none	tachycardia hypotension hypertension myocardial ischemia circulatory collapse S-T segment depression bigeminy extrasystoles ventricular fibrillation angina pectoris bradycardia transient ischemic attack thrombophlebitis
Nervous	pain (2.8%) burning sensation (1.4%)	vasovagal reaction tingling in arms grimace faintness
Digestive	nausea (1.2%)	vomiting anorexia
Respiratory	none	throat constriction dyspnea pulmonary edema
Skin and Appendages	none	rash urticaria pruritus flushing
Body as a Whole	hot flashes (1.5%)	headache fever chills excessive sweating back spasm
Special Senses	warmth (1.1%)	taste alterations nasal congestion visual disturbances
Urogenital	none	urinary retention

Regardless of the contrast agent employed, the overall estimated incidence of serious adverse reactions is higher with *coronary arteriography* than with other procedures. Cardiac decompensation, serious arrythmias, or myocardial ischemia or infarction may occur during *coronary arteriography and left ventriculography.* Following coronary and ventricular injections, certain electrocardiographic changes (increased QTc, increased R-R, T-wave amplitude) and certain hemodynamic changes (decreased systolic pressure) occurred less frequently with ISOVUE (Iopamidol Injection) than with diatrizoate meglumine and diatrizoate sodium injection; increased LVEDP occurred less frequently after ventricular iopamidol injections.

In *aortography,* the risks of procedures also include injury to the aorta and neighboring organs, pleural puncture, renal damage including infarction and acute tubular necrosis with oliguria and anuria, accidental selective filling of the right renal artery during the translumbar procedure in the presence of preexisting renal disease, retroperitoneal hemorrhage from the translumbar approach, and spinal cord injury and pathology associated with the syndrome of transverse myelitis.

Adverse effects reported in clinical literature for iopamidol include arrhythmia, arterial spasms, hematuria, periorbital edema, involuntary leg movement, malaise, and triggering of deglutition; some of these may occur as a consequence of the procedure. Other reactions may also occur with the use of any contrast agent as a consequence of the procedural hazard; these include hemorrhage or pseudoaneurysms at the puncture site, brachial plexus palsy following axillary artery injections, chest pain, myocardial infarction, and transient changes in hepatorenal chemistry tests. Arterial thrombosis, displacement of arterial plaques, venous thrombosis, dissection of the coronary vessels and transient sinus arrest are rare complications.

General Adverse Reactions to Contrast Media
Reactions known to occur with parenteral administration of iodinated ionic contrast agents (see the listing below) are possible with any nonionic agent. Approximately 95 percent of adverse reactions accompanying the use of other water-soluble intravascularly administered contrast agents are mild to moderate in degree. However, life-threatening reactions and fatalities, mostly of cardiovascular origin, have occurred. Reported incidences of death from the administration of other iodinated contrast media range from 6.6 per 1 million (0.00066 percent) to 1 in 10,000 patients (0.01 percent). Most deaths occur during injection or 5 to 10 minutes later, the main feature being cardiac arrest with cardiovascular disease as the main aggravating factor. Isolated reports of hypotensive collapse and shock are found in the literature. The incidence of shock is estimated to be 1 out of 20,000 (0.005 percent) patients.

Adverse reactions to injectable contrast media fall into two categories; chemotoxic reactions and idiosyncratic reactions. Chemotoxic reactions result from the physicochemical properties of the contrast medium, the dose, and the speed of injection. All hemodynamic disturbances and injuries to organs or vessels perfused by the contrast medium are included in this category. Experience with iopamidol suggests there is much less discomfort (e.g., pain and/or warmth) with peripheral arteriography. Fewer changes are noted in ventricular function after ventriculography and coronary arteriography.

Idiosyncratic reactions include all other reactions. They occur more frequently in patients 20 to 40 years old. Idiosyncratic reactions may or may not be dependent on the amount of drug injected, the speed of injection, the mode of injection, and the radiographic procedure. Idiosyncratic reactions are subdivided into minor, intermediate, and severe. The minor reactions are self-limited and of short duration; the severe reactions are life-threatening and treatment is urgent and mandatory.

The reported incidence of adverse reactions to contrast media in patients with a history of allergy is twice that for the general population. Patients with a history of previous reactions to a contrast medium are three times more susceptible than other patients. However, sensitivity to contrast media does not appear to increase with repeated examinations. Most adverse reactions to intravascular contrast agents appear within one to three minutes after the start of injection, but delayed reactions may occur (see PRECAUTIONS, General).

In addition to the adverse drug reactions reported for iopamidol, the following additional adverse reactions have been reported with the use of other intravascular contrast agents and are possible with the use of any water-soluble iodinated contrast agent:
Cardiovascular: vasodilation, cerebral hematomas, petechiae.
Nervous: paresthesia, dizziness, convulsions, paralysis, coma.
Respiratory: increased cough, asthma, laryngeal edema, pulmonary edema, bronchospasm, rhinitis.
Skin and Appendages: Injection site pain usually due to extravasation and/or erythematous swelling, skin necrosis.
Urogenital: osmotic nephrosis of proximal tubular cells, renal failure, pain.
Special Senses: bilateral ocular irritation; lacrimation; conjunctival chemosis, infection and conjunctivitis.

The following reactions may also occur: neutropenia, thrombophlebitis, flushing, pallor, weakness, severe retching and choking, wheezing, cramps, tremors, and sneezing.

OVERDOSAGE

Treatment of an overdose of an injectable radiopaque contrast medium is directed toward the support of all vital functions, and prompt institution of symptomatic therapy. Intravenous LD_{50} values (gl/kg) for iopamidol in animals: 21.8 (mice), 13.8 (rats), 9.6 (rabbits), 17.0 (dogs).

DOSAGE AND ADMINISTRATION

General

It is desirable that solutions of radiopaque diagnostic agents for intravascular use be at body temperature when injected. In the event that crystallization of the medium has occurred, place the vial in hot (60°-100°C) water for about five minutes, then shake gently to obtain a clear solution. Cool to body temperature before use. Discard vial without use if solids persists.

Withdrawal of contrast agents from their containers should be accomplished under aseptic conditions with sterile syringes. Sterile techniques must be used with any intravascular injection, and with catheters and guidewires.

Parenteral drug products should be inspected visually for particulate matter and discoloration prior to administration, whenever solution and container permit. Iopamidol solutions should be used only if clear and within the normal colorless to pale yellow range.

Patients should be well hydrated prior to and following ISOVUE (Iopamidol Injection) administration.

As with all radiopaque contrast agents, only the lowest dose of ISOVUE necessary to obtain adequate visualization should be used. A lower dose reduces the possibility of an adverse reaction. Most procedures do not require use of either a maximum dose or the highest available concentration of ISOVUE; the combination of dose and ISOVUE concentration to be used should be carefully individualized, and factors such as age, body size, size of the vessel and its blood flow rate, anticipated pathology and degree and extent of opacification required, structure(s) or area to be examined, disease processes affecting the patient, and equipment and technique to be employed should be considered.

Cerebral Arteriography

ISOVUE-300 (Iopamidol Injection, 300 mgl/mL) should be used. The usual individual injection by carotid puncture or transfemoral catheterization is 8 to 12 mL, with total multiple doses ranging to 90 mL.

Peripheral Arteriography

ISOVUE-300 usually provides adequate visualization. For injection into the femoral artery or subclavian artery, 5 to 40 mL may be used; for injection into the aorta for a distal runoff, 25 to 50 mL may be used. Doses up to a total of 250 mL of ISOVUE-300 have been administered during peripheral arteriography.

Peripheral Venography (Phlebography)

ISOVUE-200 (Iopamidol Injection, 200 mgl/mL) should be used. The usual dose is 25 to 150 mL per lower extremity. The combined total dose for multiple injections has not exceeded 350 mL.

Selective Visceral Arteriography and Aortography

ISOVUE-370 (Iopamidol Injection, 370 mgl/mL) should be used. Doses up to 50 mL may be required for injection into the larger vessels such as the aorta or celiac artery; doses up to 10 mL may be required for injection into the renal arteries. Often, lower doses will be sufficient. The combined total dose for multiple injections has not exceeded 225 mL.

Pediatric Angiocardiography

ISOVUE-370 should be used. Pediatric angiocardiography may be performed by injection into a large peripheral vein or by direct catheterization of the heart.

The usual dose range for single injections is provided in the following table:

Single Injection Usual Dose Range	
Age	mL
<2 years	10-15
2-9 years	15-30
10-18 years	20-50

The usual dose for cumulative injections is provided in the following table:

Cumulative Injections Usual Dose Range	
Age	mL
<2 years	40
2-4 years	50
5-9 years	100
10-18 years	125

Coronary Arteriography and Ventriculography

ISOVUE-370 should be used. The usual dose for selective coronary artery injections is 2 to 10 mL. The usual dose for ventriculography, or for nonselective opacification of multiple coronary arteries following injection at the aortic root, is 25 to 50 mL. The total dose for combined procedures has not exceeded 200 mL. EKG monitoring is essential.

Excretory Urography

ISOVUE-300 should be used. The usual adult dose is 50 mL administered by rapid intravenous injection; doses up to 100 mL have been administered.

Pediatric Excretory Urography

ISOVUE-300 should be used. The dosage recommended for use in children for excretory urography is 1.0 mL/kg to 3.0 mL/kg. It should not be necessary to exceed a total dose of 30 gl.

Computed Tomography

ISOVUE-300 may be used. *Head Imaging:* The suggested dose range is 50 to 100 mL by intravenous administration; imaging may be performed immediately after completion of administration. *Body Imaging:* The usual adult dose is 100 mL administered by rapid intravenous or bolus injection. Imaging is performed immediately after injection.

Equivalent doses of ISOVUE-370, based on organically bound iodine content, may also be used.

Pediatric Computed Tomography

ISOVUE-300 should be used. The dosage recommended for use in children for contrast enhanced computed tomography is 1.0 mL/kg to 3.0 mL/kg. It should not be necessary to exceed a total dose of 30 gl.

Drug Incompatibilities

Many radiopaque contrast agents are incompatible *in vitro* with some antihistamines and many other drugs; therefore, no other pharmaceuticals should be admixed with contrast agents.

HOW SUPPLIED

ISOVUE-200 (Iopamidol Injection 41%)
    Ten 50 mL single dose vials (NDC 0003-1314-30)
    Ten 100 mL single dose bottles (NDC 0003-1314-34)
    Ten 200 mL single dose bottles (NDC 0003-1314-40)
ISOVUE-300 (Iopamidol Injection 61%)
    Ten 50 mL single dose vials (NDC 0003-1315-30)
    Ten 100 mL single dose bottles (NDC 0003-1315-35)
    Ten 150 mL single dose bottles (NDC 0003-1315-37)
ISOVUE-370 (Iopamidol Injection 76%)
    Ten 20 mL single dose vials (NDC 0003-1316-07)
    Ten 50 mL single dose vials (NDC 0003-1316-30)
    Ten 100 mL single dose bottles (NDC 0003-1316-35)
    Ten 150 mL single dose bottles (NDC 0003-1316-37)
    Ten 200 mL single dose bottles (NDC 0003-1316-40)

Storage

Store at room temperature not exceeding 86°F. Protect from light.

Also Available

Iopamidol Injection is also available as ISOVUE-M® for intrathecal administration.

(J3-652L/R10-90)

Under license from Bracco Industria Chimica (Italy)
US Patent 4,001,323

RENOGRAFIN®-60

Diatrizoate Meglumine and Diatrizoate Sodium Injection USP

DESCRIPTION

Renografin-60 (Diatrizoate Meglumine and Diatrizoate Sodium Injection USP) is a radiopaque contrast agent supplied as a sterile, aqueous solution. Each ml provides 520 mg diatrizoate meglumine and 80 mg diatrizoate sodium; at manufacture, 3.2 mg sodium citrate and 0.4 mg edetate disodium are added per ml. The pH has been adjusted between 6.0 and 7.7 with sodium hydroxide and diatrizoic acid. *Each ml of solution also contains approximately 3.76 mg (0.16 mEq) sodium* and 292.5 mg organically bound iodine. At the time of manufacture, the air in the container is replaced by nitrogen.

CLINICAL PHARMACOLOGY

Following intravascular injection, Renografin-60 is rapidly transported through the bloodstream to the kidneys and is excreted unchanged in the urine by glomerular filtration. When urinary tract obstruction is severe enough to block glomerular filtration, the agent appears to be excreted by the tubular epithelium.

Certain applications of the contrast agent make use of the natural physiologic mechanism of excretion. Thus, the intravenous injection of the agent permits visualization of the kidneys and urinary passages.

Renal accumulation is sufficiently rapid that the period of maximal opacification of the renal passages may begin as early as five minutes after injection. In infants and small children excretion takes place somewhat more promptly than in adults, so that maximal opacification occurs more rapidly and is less sustained. The normal kidney eliminates the contrast medium almost immediately. In nephropathic conditions, particularly when excretory capacity has been altered, the rate of excretion varies unpredictably, and opacification may be delayed for 30 minutes or more after injection; with severe impairment opacification may not occur. Generally, however, the medium is concentrated in sufficient amounts and promptly enough to permit a thorough evaluation of the anatomy and physiology of the urinary tract. After intramuscular injection, the contrast agent is promptly absorbed and normally reaches the renal passages within 20 to 60 minutes.

Intravascular injection of diatrizoate also opacifies those vessels in the path of flow of the medium, permitting visualization until the circulating blood dilutes the concentration of the medium. Thus selective angiography may be performed following injection directly into veins or arteries such as the carotid, the vertebral, or the vessels of the extremities.

Under certain circumstances, specific parts of the body which do not concentrate the contrast agent physiologically may be visualized by injecting the agent directly into the region to be studied. The biliary tract is one organ system which may be visualized in this manner. In operative cholangiography, injection of the radiopaque medium into the cystic duct or choledochal lumen, at laparotomy, opacifies the intra- and extra-hepatic biliary ductal system, revealing the nature and location of obstructions such as stones or strictures. Injection of the medium through an in-place T-tube, immediately after exploration of the common duct, permits the visualization of retained stones. A repetition of "T-tube cholangiography," performed as part of the postoperative follow-up, insures the patency of the ductal system before removal of the T-tube. The biliary ductal system may also be opacified by the percutaneous-transhepatic route. In relatively long-standing biliary obstruction, the biliary ducts are usually enlarged sufficiently to be located promptly by percutaneous transhepatic probing, permitting injection of the contrast agent directly into the biliary ductal system.

If the contrast agent is injected directly into the splenic pulp, significant opacification of the splenic and portal veins is obtained. Because of gravity, the dependent portions of the portal system are better opacified than the superior portions. The agent is carried from the portal vein into the hepatic veins, and a diffuse opacification of the liver results. in patients with portal hypertension, collateral pathways caused by the change in portal blood flow may be visualized and esophageal varices are often delineated. The procedure may reveal the site of portal obstruction.

Injection of Renografin-60 directly into a joint space provides visual information about joint derangements.

A small amount of the radiopaque agent injected into a normal cervical or lumbar disk will, under optimal conditions, concentrate within the nucleus pulposus. In the presence of disk pathology the injected agent may reveal significant bulging or disruption of the annulus beyond its normal confines and may identify disk degeneration, retropulsion, or rupture.

Computed Tomography

Renografin-60 enhances computed tomographic brain scanning through augmentation of radiographic efficiency. The degree of enhancement of visualization of tissue density is directly related to the iodine content in an administered dose; peak iodine blood levels occur immediately following rapid injection of the dose. These levels fall rapidly within five to ten minutes. This can be accounted for by the dilution in the vascular and extracellular fluid compartments which causes an initial sharp fall in plasma concentration. Equilibration with the extracellular compartments is reached in about ten minutes; thereafter, the fall becomes exponential. Maximum contrast enhancement frequently occurs after peak blood iodine levels are reached. The delay in maximum contrast enhancement can range from five to forty minutes, depending on the peak iodine levels achieved and the cell type of the lesion. This lag suggests that radiographic contrast enhancement is at least in part dependent on the accumulation of iodine within the lesion and outside the blood pool, although the mechanism by which this occurs is not clear. The radiographic enhancement of nontumoral lesions, such as arteriovenous malformations and aneurysms is probably dependent on the iodine content of the circulating blood pool.

In brain scanning, Renografin-60 (Diatrizoate Meglumine and Diatrizoate Sodium Injection USP) does not accumulate in normal brain tissue due to the presence of the "blood-brain" barrier. The increase in X-ray absorption in normal brain is due to the presence of contrast agent within the blood pool. A break in the blood-brain barrier such as occurs in malignant tumors of the brain allows the accumulation of the contrast medium within the interstitial tumor tissue. Adjacent normal brain tissue does not contain the contrast medium.

In nonneural tissues (during computed tomography of the body), diatrizoate diffuses rapidly from the vascular into the extravascular space. Increase in X-ray absorption is related to blood flow, concentration of the contrast medium, and extraction of the contrast medium by interstitial tumor tissue since no barrier exists. Contrast enhancement is thus due to the relative differences in extravascular diffusion between normal and abnormal tissue, quite different from that in the brain.

The pharmacokinetics of diatrizoate in both normal and abnormal tissue have been shown to be variable. Contrast enhancement appears to be greatest soon after administration of the contrast medium, and following intra-arterial rather than intravenous administration. Thus, greatest enhancement can be detected by a series of consecutive two- to three-second scans performed just after injection (within 30 to 90 seconds), i.e., dynamic computed tomographic scanning.

INDICATIONS

Renografin-60 is indicated in excretion urography (by direct I.V. or drip infusion); cerebral angiography; peripheral arteriography; venography; operative, T-tube, or percutaneous transhepatic cholangiography; splenoportography; arthrography; and discography.

Computed Tomography

Renografin-60 is also indicated for radiographic contrast enhancement in computed tomography (CT) of the brain and body. Contrast enhancement may be advantageous in delineating or ruling out disease in suspicious areas which may otherwise not have been satisfactorily visualized.

Brain Tumors

Renografin-60 may be useful to demonstrate the presence and extent of certain malignancies such as: gliomas including malignant gliomas, glioblastomas, astrocytomas, oligodendrogliomas and gangliomas; ependymomas; medulloblastomas; meningiomas; neuromas; pinealomas; pituitary adenomas; craniopharyngiomas; germinomas; and metastatic lesions.

The usefulness of contrast enhancement for the investigation of the retrobulbar space and in cases of low grade or infiltrative glioma has not been demonstrated. In cases where lesions have calcified, there is less likelihood of enhancement. Following therapy, tumors may show decreased or no enhancement.

Non-Neoplastic Conditions of the Brain
The use of Renografin-60 may be beneficial in the enhancement of images of lesions not due to neoplasms. Cerebral infarctions of recent onset may be better visualized with the contrast enhancement, while some infarctions are obscured if a contrast medium is used. The use of Renografin-60 improved the contrast enhancement in approximately 60 percent of cerebral infarctions studied from one week to four weeks from the onset of symptoms.

Sites of active infection also will produce contrast enhancement following contrast medium administration.

Arteriovenous malformations and aneurysms will show contrast enhancement. In the case of these vascular lesions, the enhancement is probably dependent on the iodine content of the circulating blood pool.

Hematomas and intraparenchymal bleeders seldom demonstrate any contrast enhancement. However, in cases of intraparenchymal clot, for which there is no obvious clinical explanation, contrast medium administration may be helpful in ruling out the possibility of associated arteriovenous malformation.

The opacification of the inferior vermis following contrast medium administration has resulted in false-positive diagnoses in a number of normal studies.

Body Scanning
Renographin-60 (Diatrizoate Meglumine and Diatrizoate Sodium Injection USP) may be used for enhancement of computed tomographic scans performed for detection and evaluation of lesions in the liver, pancreas, kidneys, aorta, mediastinum, abdominal cavity, pelvis and retroperitoneal space.

Enhancement of computed tomography with Renografin-60 may be of benefit in establishing diagnoses of certain lesions in these sites with greater assurance than is possible with CT alone, and in supplying additional features of the lesions (e.g. hepatic abscess delineation prior to percutaneous drainage). In other cases, the contrast agent may allow visualization of lesions not seen with CT alone (e.g., tumor extension), or may help to define suspicious lesions seen with unenhanced CT (e.g., pancreatic cyst).

Contrast enhancement appears to be greatest within 60-90 seconds after bolus administration of the contrast agent. Therefore, utilization of a continuous scanning technique ("dynamic CT scanning") may improve enhancement and diagnostic assessment of tumor and other lesions such as an abscess, occasionally revealing unsuspected or more extensive disease. For example, a cyst can be distinguished from a vascularized solid lesion when pre-contrast and enhanced scans are compared; the non-perfused mass shows are unchanged X-ray absorption (CT number). A vascularized lesion is characterized by an increase in CT number in the few minutes after a bolus of intravascular contrast agent; it may be malignant, benign or normal tissue, but would probably not be a cyst, hematoma, or other nonvascular lesion.

Because unenhanced scanning may provide adequate diagnostic information in the individual patient, the decision to employ contrast enhancement, which may be associated with risk and increased radiation exposure, should be based upon a careful evaluation of clinical, other radiological, and unenhanced CT findings.

CONTRAINDICATIONS
This preparation is contraindicated in patients with a hypersensitivity to salts of diatrizoic acid.

Urography is contraindicated in patients with anuria.

Specific contraindication to **percutaneous transhepatic cholangiography** include a prothrombin time below 50 percent and evidence of coagulation defects.

Splenoportography should not be performed on any patient for whom splenectomy is contraindicated, since complications of the procedure at times make splenectomy necessary. Other contraindications include prothrombin time below 50 percent, signif-

icant thrombocytopenia or coagulation defect, and any condition which may increase the possibility of rupture of the spleen.

Arthrography should not be performed infection is present in or near the joint.

Discography should not be performed in patients with an infection or open injury near the region to be examined.

WARNINGS
Ionic iodinated contrast media inhibit blood coagulation, *in vitro*, more than nonionic contrast media. Nonetheless, it is prudent to avoid prolonged contact of blood with syringes containing ionic contrast media.

Serious, rarely fatal, thromboembolic events causing myocardial infarction and stroke have been reported during angiographic procedures with both ionic and nonionic contrast media. Therefore, meticulous intravascular administration technique is necessary, particularly during angiographic procedures, to minimize thromboembolic events. Numerous factors, including length of procedure, catheter and syringe material, underlying disease state, and concomitant medications may contribute to the development of thromboembolic events. For these reasons, meticulous angiographic techniques are recommended including close attention to guidewire and catheter manipulation, use of manifold systems and/or three way stopcocks, frequent catheter flushing with heparinized saline solutions, and minimizing the length of the procedure. The use of plastic syringes in place of glass syringes has been reported to decrease but not eliminate the likelihood of *in vitro* clotting.

A definite risk exists in the use of intravascular contrast agents in patients who are known to have multiple myeloma. In such instances there has been anuria resulting in progressive uremia, renal failure, and eventually death. Although neither the contrast agent nor dehydration has separately proved to be the cause of anuria in myeloma, it has been speculated that the combination of both may be the causative factor. The risk in myelomatous patients is not a contraindication to the procedures; however, partial dehydration in the preparation of these patients for the examination is not recommended since this may predispose to the precipitation of myeloma protein in the renal tubules. No form of therapy, including dialysis, has been successful in reversing this effect. Myeloma, which occurs most commonly in persons over age 40, should be considered before intravascular administration of a contrast agent.

Administration of radiopaque materials to patients known or suspected to have pheochromocytoma should be performed with extreme caution. If, in the opinion of the physician, the possible benefits of such procedures outweigh the considered risks, the procedures may be performed; however, the amount of radiopaque medium injected should be kept to an absolute minimum. The blood pressure should be assessed throughout the procedure and measures for treatment of a hypertensive crisis should be available.

Contrast media have been shown to promote the phenomenon of sickling in individuals who are homozygous for sickly cell disease when the material is injected intravenously or intra-arterially.

Since iodine-containing contrast agents may alter the results of thyroid function tests, such tests, if indicated, should be performed prior to the administration of this preparation.

A history of sensitivity to iodine *per se* or to other contrast agents is not an absolute contraindication to the use of diatrizoate but calls for extreme caution in administration.

Avoid accidental introduction of this preparation into the subarachnoid space since even small amounts may produce convulsions and possible fatal reactions. In patients with subarachnoid hemorrhage, a rare association between contrast administration and clinical deterioration, including convulsions and death, has been reported; therefore, administration of intravascular iodinated ionic contrast media in these patients should be undertaken with caution.

Cerebral angiography should be undertaken with special caution in extreme age, poor clinical condition, advanced arteriosclerosis, severe arterial hypertension, cardiac decompensation, recent cerebral embolism, or thrombosis.

Urography should be performed with extreme caution in patients with severe concomitant hepatic and renal disease.

PRECAUTIONS

Diagnostic procedures which involve the use of radiopaque contrast agents should be carried out under the direction of personnel with the prerequisite training and with a thorough knowledge of the particular procedure to be performed (see ADVERSE REACTIONS).

Severe, life-threatening reactions suggest hypersensitivity to the radiopaque agent, which has prompted the use of several pretesting methods, none of which can be relied upon to predict severe reactions. Many authorities question the value of any pretest. A history of bronchial asthma or allergy, a family history of allergy, or a previous reaction to a contrast agent warrant special attention. Such a history, by suggesting histamine sensitivity and a consequent proneness to reactions, may be more accurate than pretesting in predicting the likelihood of a reaction, although not necessarily the severity or type of reaction in the individual case.

The sensitivity test most often performed is the slow injection of 0.5 to 1.0 ml of the radiopaque medium, administered intravenously, prior to injection of the full diagnostic dose. It should be noted that the absence of a reaction to the test dose does not preclude the possibility of a reaction to the full diagnostic dose. If the test dose causes an untoward response of any kind, the necessity for continuing with the examination should be carefully reevaluated and, if it is deemed essential, the examination should be conducted with all possible caution. In rare instances, reactions to the test dose itself may be extremely severe; therefore, close observation of the patient, and facilities for emergency treatment, appear indicated.

Renal toxicity has been reported in a few patients with liver dysfunction who were given oral cholecystographic agents followed by urographic agents. Administration of Renografin-60 (Diatrizoate Meglumine and Diatrizoate Sodium Injection USP) should therefore be postponed in any patient with a known or suspected hepatic or biliary disorder who has recently taken a cholecystographic contrast agent.

Caution should be exercised with the use of radiopaque media in severely debilitated patients and in those with marked hypertension. The possibility of thrombosis should be borne in mind when percutaneous techniques are employed.

Consideration must be given to the functional ability of the kidneys before injecting this preparation.

Contrast agents may interfere with some chemical determinations made on urine specimens; therefore, urine should be collected before administration of the contrast medium or two or more days afterwards.

The following precautions pertain to specific procedures:

Peripheral arteriography: Hypotension or moderate decreases in blood pressure seem to occur frequently with intra-arterial (brachial) injections; therefore, the blood pressure should be monitored during the immediate ten minutes after injection; this blood pressure change is transient and usually requires no treatment.

Excretion urography: Adequate visualization may be difficult or impossible to attain in uremic patients or others with severely impaired renal function (see CONTRAINDICATIONS). The increased osmotic load associated with drip infusion pyelography should be considered in patients with congestive heart failure. The diuretic effect of the drip infusion pyelography procedure may hinder assessment of residual urine in the bladder. The recommended rate of infusion should not be exceeded.

Acute renal failure has been reported in diabetic patients with diabetic nephropathy and susceptible nondiabetic patients (often elderly with preexisting renal disease) following excretion urography. Therefore, careful consideration should be given before performing this procedure in these patients.

Operative and T-tube cholangiography: Injection should be made slowly to prevent extravasation of the medium into the peritoneal cavity, and to minimize reflux flow into the pancreatic duct which may result in pancreatic irritation.

Percutaneous transhepatic cholangiography: To reduce the possibility of bile leakage and consequent peritonitis, as much of the contrast agent as possible should be aspirated on completion of successful films. All patients should be carefully and constantly monitored for 24 hours after the procedure for signs of internal hemorrhage or bile leakage; if these complications are recognized immediately, remedial measures can be instituted promptly with minimal increase in morbidity. Percutaneous transhepatic cholangiography is not without risk and should therefore be reserved for special circumstances when ordinary studies of the biliary system have failed to provide the requisite information in jaundiced patients who are not good candidates for surgery. The procedure should only be attempted when competent surgical intervention can be promptly obtained if needed.

Splenoportography: It is best to avoid manipulations which would prolong the time the needle is in the spleen, since they may contribute to subcapsular extravasation of the contrast agent, and also to postpuncture bleeding. Following splenoportography, the patient should lie on his left side for several hours and should be closely observed for 24 hours for signs of internal bleeding, which is the most common complication of the procedure. Fatal hemorrhage has occurred on rare occasion, but leakage of up to 300 ml of blood from the spleen is apparently not uncommon. Blood transfusions may be required, and rarely splenectomy.

Discography: To minimize the possibility of introducing infection, discography should be postponed in any patient with an infection or open injury near the region to be examined, including upper respiratory infections in the case of cervical discography. All possible care should be taken to preclude contamination and resultant infection of the disk, which has been reported after discography. In cervical discography, particular care is needed to avoid puncturing the esophagus and thereby introducing contamination into the disk. Rupture of the disk is highly unlikely if care in performance is observed, but may occur if the point of the needle has been barbed by contact with bone; use of the two-needle technique should help reduce this hazard.

USAGE IN PREGNANCY

Safety for use during pregnancy has not been established; therefore, this preparation should be used in pregnant patients only when, in the judgment of the physician, its use is deemed essential to the welfare of the patient.

ADVERSE REACTIONS

Adverse reactions accompanying the use of iodine-containing intravascular contrast agents are usually mild and transient although severe and life-threatening reactions, including fatalities, have occurred. Because of the possibility of severe reactions to the procedure and/or the radiopaque medium, appropriate emergency facilities and well-trained personnel should be available to treat both conditions. Emergency facilities and personnel should remain available for 30 to 60 minutes following the procedure since severe delayed reactions have been known to occur.

Nausea, vomiting, flushing, or a generalized feeling of warmth are the reactions seen most frequently with intravascular injection. Symptoms which may occur are chills, fever, sweating, headache, dizziness, pallor, weakness, severe retching and choking, wheezing, a rise or fall in blood pressure, facial or conjunctival petechiae, urticaria, pruritus, rash, and other eruptions, edema, cramps, tremors, itching, sneezing, lacrimation, etc. Antihistaminic agents may be of benefit; rarely such reactions may be severe enough to require discontinuation of dosage.

Severe reactions which may require emergency measures may take the form of a cardiovascular reaction characterized by peripheral vasodilatation with resultant hypotension and reflex tachycardia, dyspnea, agitation, confusion and cyanosis progressing to unconsciousness. Or, the histamine-liberating effect of these compounds may induce an allergic-like reaction which may range in severity from rhinitis or angioneurotic edema to laryngeal or bronchial spasm or anaphylactoid shock.

Temporary renal shutdown or other nephropathy may occur. Temporary neurologic effects of varying severity have occurred in a few instances, particularly when the medium was used for angiography in the diagnosis of cerebral pathology. Although local tissue tolerance is usually good, there have been a few reports of a burning or stinging sensation or numbness and of venospasm or venous pain, and partial collapse of the injected vein. Neutropenia or thrombophlebitis may occur.

Adverse effects may sometimes occur as a consequence of the procedure for which the contrast agent is used. Adverse reactions in **excretion urography** have included cardiac arrest, ventricular fibrillation, anaphylaxis with severe asthmatic reaction, and flushing due to generalized vasodilation. **Cerebral angiography** has been known to cause temporary neurologic complications such as induction of seizures, particularly in patients with convulsive disorders; confusional states or drowsiness; transient paresis; coma; temporary disturbances in vision; or seventh nerve weakness. During **peripheral arteriography**, complications have occurred including hemorrhage from the puncture site, thrombosis of the vessel, and brachial plexus palsy following axillary artery injections.

Complications of **percutaneous transhepatic cholangiography** have been estimated to occur in four to six percent of cases and have included bile leakage and biliary peritonitis, gallbladder perforation, internal bleeding, septicemia involving gram-negative organisms, and tension pneumothorax from inadvertent puncture of the diaphragm and lung. Bile leakage may be more likely in patients with complete obstruction due to carcinoma.

During **splenoportography,** intraperitoneal extravasation of the contrast medium may cause transient diaphragmatic irritation or mild to moderate transient pain which may sometimes be referred to the shoulder, the periumbilical region, or other areas. Because of the proximity of the pleural cavity, accidental pneumothorax has been known to occur. Inadvertent injection of the medium into other nearby structures is not likely to cause untoward consequences.

Arthrography may induce joint pain or increase existing pain, particularly if a large dose is used and the medium extravasates into surrounding soft tissue. Pain or discomfort is usually immediate and transient but may be delayed or of extended duration (up to 24 hours). Lipid-filled histiocytes have been found in tissue removed following arthrography. The technique of **discography** may be painful, particularly when disk pathology exists. Pain on injection may also be related to the volume of the dose. The nature of the disk pathology or extravasation of contrast agent may cause referred pain.

When any percutaneous technique is employed the possibility of thrombosis or of other complications due to the mechanical trauma of the procedure should be borne in mind.

DOSAGE AND ADMINISTRATION

Renografin-60 (Diatrizoate Meglumine and Diatrizoate Sodium Injection USP) should be at body temperature when injected, and may need to be warmed before use. If kept in a syringe for prolonged periods before injection, it should be protected from exposure to strong light.

Dilution and withdrawal of the contrast agent should be accomplished under aseptic conditions with sterile needle and syringe.

Excretion Urography

Appropriate preparation of the patient is desirable for optimal results. In adults and older children, a laxative the night before the examination, a low residue diet the day before, and low liquid intake for 12 hours prior to the procedure may be used to clear the gastrointestinal tract and to induce a partial dehydration which is believed to increase the urinary concentration of the contrast medium. Preparatory partial dehydration is not recommended in infants, young children, the elderly or azotemic patients (especially those with polyuria, oliguria, diabetes, advanced vascular disease, or preexisting dehydration). The undesirable dehydration in these patients may be accentuated by the osmotic diuretic action of the medium.

In uremic patients partial dehydration is not necessary and maintenance of adequate fluid intake is particularly desirable.

Direct I.V. Injection: The dose range for adults is 25 to 50 ml; the usual dose is 25 ml; children require proportionately less. Suggested dosages are as follows: under 6 months – 5 ml; 6 to 12 months – 8 ml;1 to 2 years – 10 ml; 2 to 5 years – 12 ml; 5 to 7 years – 15 ml; 8 to 10 years – 18 ml; 11 to 15 years – 20 ml; adults (16 years and older) – 25 to 50 ml. In adults, when the smaller dose has provided inadequate visualization, or when poor visualization is anticipated, the 50 ml dose may be given. Drip infusion may be used when direct I.V. pyelography is not expected to be or has not been satisfactory (see below).

The preparation is given by intravenous injection. If flushing or nausea occurs during administration, injection should be slowed or briefly interrupted until the side effects have disappeared.

A scout film should be made before the contrast medium is administered. To allow for individual variation, several films should be exposed beginning approximately five minutes after injection. In patients with renal dysfunction optimal visualization may be delayed until 30 minutes or more after injection.

NOTE: In infants and children and in certain adults the medium may be injected intramuscularly. The suggested dose is 25 ml for adults and proportionately less for children, divided and given bilaterally in the gluteal muscles. Radiographs should be taken at 20, 40 and 60 minutes after the medium is injected.

Drip Infusion Pyelography: In drip infusion pyelography, the recommended dose of Renografin-60 (Diatrizoate Meglumine and Diatrizoate Sodium Injection USP) is calculated on the basis of 1 ml of Renografin-60 per pound of body weight diluted with an equal volume of Sterile Water for Injection USP. The diluted preparation (30%) is given by I.V. infusion through a large bore (17- to 18-gauge) needle at a rate of 40 ml per minute. The recommended rate of infusion should not be exceeded and the total volume administered should generally not exceed 300 ml. In older patients and in patients with known or suspected cardiac decompensation, a slower rate of infusion is probably wise.

If nausea or flushing occurs during administration, the infusion should be slowed or briefly interrupted.

Films are taken before the onset of the infusion and at the desired intervals following its completion. When renal function is normal, a nephrogram may be taken as soon as the infusion is completed, and films of the collecting system at 10 and 20 minutes thereafter. Voiding cystourethrograms are usually optimal at 20 minutes after the infusion is completed. In hypertensive patients, early minute sequence films may be taken during the course of infusion, in addition to subsequent pyelograms. In patients with renal dysfunction, optimal visualization is usually delayed, and late films are taken as indicated.

The nephrogram obtained by the drip infusion procedure may be dense enough to obscure the pelvocalyceal system in some cases. The presence of gas in the bowel may hamper early visualization of the renal collecting system. Tomographic "cuts" may help to overcome such difficulties.

Nephrotomography may begin when the infusion is completed. The sustained contrast achieved by the drip infusion technique eliminates the need for precise timing and teamwork that is necessary with ordinary nephrotomography. Thus, if nephrograms taken after infusion of the medium suggest the need for sectional films, or if preselected tomographic "cuts" are not sufficient, additional tomograms may be obtained at once, and without repetition of dosage.

Cerebral Angiography

Appropriate preparation of the patient is indicated, including suitable premedication. The average single dose for adults in 10 ml, repeated as indicated. Children require less in proportion to weight.

Either the percutaneous or operative method of administration may be used. For visualization of the cerebral vessels, the contrast medium is injected into the common carotid artery; for angiography of the vessels in the posterior fossa or the occipital lobes, the medium is injected into the vertebral artery. Since the medium is given by rapid injection, the patient should be watched for untoward reactions. Unless general anesthesia is used, patients should be warned that the medium may provoke movement and that they may feel transient pain, flushing, or burning during the injection.

A scout film should be made routinely before the contrast medium is injected. Serial films begun while the last few ml are being injected should permit visualization of the arterial, intermediate, and venous phases.

Peripheral Arteriography

Appropriate preparation of the patient is indicated, including suitable premedication. For visualization of an entire extremity, a single dose of 20 to 40 ml is suggested; for the upper or lower half of the extremity only, 10 to 20 ml is usually sufficient.

Injection is made into the femoral or subclavian artery by the percutaneous or operative method. Because the contrast agent is given by rapid injection, flushing of the skin may occur. Patients not under general anesthesia may experience nausea and vomiting or transient feeling of warmth. Vascular spasm is not likely to occur.

A scout film should be made routinely before administering the contrast medium. Radiograms of the upper half of the extremity are taken while the last few ml are being injected, followed by radiograms of the lower half of the extremity a few seconds later.

Venography

For visualization of veins in the upper extremities, a single dose of 10 ml per extremity is suggested. For veins in the lower extremities, doses of 20 to 40 ml per extremity are suggested. In exceptional circumstances, larger doses may be necessary; visualization of the iliac vein, extensive varicosities or large veins may require 50 ml or more. Total doses up to 100 ml per lower extremity have been used safely.

For visualization of an upper extremity, the medium may be given by percutaneous injection into any convenient superficial vein of the forearm or hand. For the visualization of a lower extremity it should be injected into a superficial vein on the lateral side of the foot. The medium is injected rapidly; patients should be observed for untoward reactions.

Radiograms are taken when injection is completed; sufficient time should be allowed to permit diffusion of the contrast medium.

Operative and Postoperative Cholangiography

Operative cholangiography is performed as soon as the gallbladder and ducts have been exposed surgically. The usual dose is 10 ml but as much as 25 ml may be needed, depending on the caliber of the ducts. If desired, the contrast agent may be diluted 1:1 with Sodium Chloride Injection USP under strict aseptic procedures.

The contrast medium is instilled slowly through the stump of the cystic duct or directly into the choledochal lumen. Following surgical exploration of the ductal system, repeat studies may be performed before closure of the abdomen, using the same dose as before.

Postoperatively, the ductal system may be examined by injection of the contrast agent through an in-place T-tube. "T-tube cholangiography" is usually performed eight to ten days after operation; the usual dose is the same as for operative cholangiography.

For each procedure, films are taken immediately after instillation of the medium and are read immediately. Additional films are then taken if necessary.

Percutaneous Transhepatic Cholangiography

Facilities for emergency surgery should be available whenever this examination is performed. Appropriate premedication of the patient is recommended; drugs which are likely to cause spasm, such as morphine, should be avoided.

Depending on the caliber of the biliary tree, a dose of 20 to 40 ml is generally sufficient to opacify the entire ductal system. The contrast agent may be diluted 1:1 with Sodium Chloride Injection USP, if desired, under strict aseptic procedures.

Injection is made into a biliary duct by the percutaneous transhepatic method. Before the dose is administered, as much bile as possible is aspirated. The medium is then slowly injected into the duct under very slight pressure. If a duct is not located promptly, successive small doses of 1 to 2 ml are injected into the liver as the needle is gradually withdrawn, until a duct is visualized by x-ray. If no duct can be located after three or four attempts, the procedure is abandoned.

Serial films are taken rapidly during and after injection of the medium into the biliary ducts. Repositioning of the patient, if necessary, should be done with care.

In hepatocellular disease, the biliary ducts are generally not enlarged and cannot successfully be opacified by this method. Thus, in the presence of longstanding jaundice, failure to obtain a successful percutaneous transhepatic cholangiogram by a person experienced in the technique is generally considered to be strongly suggestive of nonobstructive or hepatocellular-type jaundice.

Splenoportography

Prior gastrointestinal x-ray examination should include particular attention to the lower esophageal are. A hematologic survey, including prothrombin time and platelet count, should be performed. The patient should have no food for several hours and should be mildly sedated. Splenoportography is usually performed under local anesthesia.

Approximately 20 to 25 ml of the contrast agent is usually adequate. The dose is injected rapidly, following radiologic location and percutaneous puncture of the spleen.

Preliminary films are taken to locate the spleen before the injection is begun. Rapid serial films are then started simultaneously with injection of the dose. Serial films are necessary since the entire portal system cannot be captured on a single film and also because of individual variations in portal circulation time.

Arthrography

The amount of contrast agent required is dependent on the size of the joint to be injected. For an adult, the following doses are generally suitable: Knee – 5 to 15 ml; shoulder or hip – 5 to 10 ml; other joints – 1 to 4 ml. Dosage for children should be suitably reduced.

The injection site should be prepared aseptically. Excessive synovial fluid should be aspirated to minimize pain and to reduce intra-articular dilution of the contrast agent. If indicated, the agent may be administered under local anesthesia. After injection of the medium, the joint should be manipulated gently in order to spread the medium throughout the joint space. In some instances, double contrast arthrography, injecting both air and contrast medium, has been of value.

Films are taken from several angles; stereoscopic films may be advantageous.

When the contrast agent is used to opacify a joint space, much of the agent may be aspirated at the end of the procedure.

Discography

No prior preparation of the patient is required, although administration of an analgesic or sedative 20 minutes before the procedure may be helpful. Discography is performed under local anesthesia using the usual aseptic precautions.

Dosage is generally determined by the amount of contrast agent which can easily be injected into the disk without force. A cervical disk will normally accept up to 0.5 ml and a lumbar disk 1 or 2 ml. The amount may vary, and injection should be discontinued when resistance is felt. The rate of injection may influence the amount which can be injected. To reduce the probability of extravasation and to minimize unnecessary pain, injection should be made slowly and not more than 2 ml should be injected into any one disk.

A two-needle technique may be used to administer the contrast medium, with a large-gauge needle to locate the disk and a small-gauge needle within the larger one to puncture the disk and administer the medium. The correct position of the two needles is established radiologically before the medium is injected.

Spot roentgenograms should be taken anteroposteriorly, obliquely, and laterally as soon as disks have been injected.

When the contrast agent is used for discography, it need not be aspirated at the end of the procedure.

Computed Tomography

Brain Scanning

The suggested dose range is 50 to 150 ml by intravenous administration; scanning may be performed immediately after completion of administration. Doses for children should be proportionately less, depending on age and weight.

Body Scanning

The usual adult dose is 100 ml administered by rapid intravenous (within approximately 1 minute) bolus injection. Scanning is performed immediately after injection.

Gastrografin® (Diatrizoate Meglumine and Diatrizoate Sodium Solution USP), an oral radiopaque contrast agent, may be useful as an adjunct to the procedure.

Patient Preparation

No special patient preparation is required for contrast enhancement of CT brain scanning or body scanning. However, it is advisable to insure that patients are well hydrated prior to examination.

HOW SUPPLIED

Renografin-60 (Diatrizoate Meglumine and Diatrizoate Sodium Injection USP) is available in 10 ml, 30 ml, 50 ml, and 100 ml single dose vials, and 100 ml single dose bottles.

STORAGE

The preparation should be stored at room temperature, protected from light. If precipitation or solidification has occurred due to storage in the cold, immerse the container in hot water and shake intermittently to redissolve any solids. (J3-441L/R10-90)